Mathematical Programming with Data Perturbations II

PURE AND APPLIED MATHEMATICS

A Program of Monographs, Textbooks, and Lecture Notes

Contributions to *Lecture Notes in Pure and Applied Mathematics* are reproduced by direct photography of the author's typewritten manuscript. Potential authors are advised to submit preliminary manuscripts for review purposes. After acceptance, the author is responsible for preparing the final manuscript in camera-ready form, suitable for direct reproduction. Marcel Dekker, Inc. will furnish instructions to authors and special typing paper. Sample pages are reviewed and returned with our suggestions to assure quality control and the most attractive rendering of your manuscript. The publisher will also be happy to supervise and assist in all stages of the preparation of your camera-ready manuscript.

LECTURE NOTES
IN PURE AND APPLIED MATHEMATICS

1. *N. Jacobson*, Exceptional Lie Algebras
2. *L.-Å. Lindahl and F. Poulsen*, Thin Sets in Harmonic Analysis
3. *I. Satake*, Classification Theory of Semi-Simple Algebraic Groups
4. *F. Hirzebruch, W. D. Newmann, and S. S. Koh*, Differentiable Manifolds and Quadratic Forms (out of print)
5. *I. Chavel*, Riemannian Symmetric Spaces of Rank One (out of print)
6. *R. B. Burckel*, Characterization of C(X) Among Its Subalgebras
7. *B. R. McDonald, A. R. Magid, and K. C. Smith*, Ring Theory: Proceedings of the Oklahoma Conference
8. *Y.-T. Siu*, Techniques of Extension of Analytic Objects
9. *S. R. Caradus, W. E. Pfaffenberger, and B. Yood*, Calkin Algebras and Algebras of Operators on Banach Spaces
10. *E. O. Roxin, P.-T. Liu, and R. L. Sternberg*, Differential Games and Control Theory
11. *M. Orzech and C. Small*, The Brauer Group of Commutative Rings
12. *S. Thomeier*, Topology and Its Applications
13. *J. M. Lopez and K. A. Ross*, Sidon Sets
14. *W. W. Comfort and S. Negrepontis*, Continuous Pseudometrics
15. *K. McKennon and J. M. Robertson*, Locally Convex Spaces
16. *M. Carmeli and S. Malin*, Representations of the Rotation and Lorentz Groups: An Introduction
17. *G. B. Seligman*, Rational Methods in Lie Algebras
18. *D. G. de Figueiredo*, Functional Analysis: Proceedings of the Brazilian Mathematical Society Symposium
19. *L. Cesari, R. Kannan, and J. D. Schuur*, Nonlinear Functional Analysis and Differential Equations: Proceedings of the Michigan State University Conference
20. *J. J. Schäffer*, Geometry of Spheres in Normed Spaces
21. *K. Yano and M. Kon*, Anti-Invariant Submanifolds
22. *W. V. Vasconcelos*, The Rings of Dimension Two
23. *R. E. Chandler*, Hausdorff Compactifications
24. *S. P. Franklin and B. V. S. Thomas*, Topology: Proceedings of the Memphis State University Conference
25. *S. K. Jain*, Ring Theory: Proceedings of the Ohio University Conference
26. *B. R. McDonald and R. A. Morris*, Ring Theory II: Proceedings of the Second Oklahoma Conference
27. *R. B. Mura and A. Rhemtulla*, Orderable Groups
28. *J. R. Graef*, Stability of Dynamical Systems: Theory and Applications
29. *H.-C. Wang*, Homogeneous Banach Algebras
30. *E. O. Roxin, P.-T. Liu, and R. L. Sternberg*, Differential Games and Control Theory II
31. *R. D. Porter*, Introduction to Fibre Bundles
32. *M. Altman*, Contractors and Contractor Directions Theory and Applications
33. *J. S. Golan*, Decomposition and Dimension in Module Categories
34. *G. Fairweather*, Finite Element Galerkin Methods for Differential Equations
35. *J. D. Sally*, Numbers of Generators of Ideals in Local Rings
36. *S. S. Miller*, Complex Analysis: Proceedings of the S.U.N.Y. Brockport Conference
37. *R. Gordon*, Representation Theory of Algebras: Proceedings of the Philadelphia Conference
38. *M. Goto and F. D. Grosshans*, Semisimple Lie Algebras
39. *A. I. Arruda, N. C. A. da Costa, and R. Chuaqui*, Mathematical Logic: Proceedings of the First Brazilian Conference

40. *F. Van Oystaeyen,* Ring Theory: Proceedings of the 1977 Antwerp Conference
41. *F. Van Oystaeyen and A. Verschoren,* Reflectors and Localization: Application to Sheaf Theory
42. *M. Satyanarayana,* Positively Ordered Semigroups
43. *D. L. Russell,* Mathematics of Finite-Dimensional Control Systems
44. *P.-T. Liu and E. Roxin,* Differential Games and Control Theory III: Proceedings of the Third Kingston Conference, Part A
45. *A. Geramita and J. Seberry,* Orthogonal Designs: Quadratic Forms and Hadamard Matrices
46. *J. Cigler, V. Losert, and P. Michor,* Banach Modules and Functors on Categories of Banach Spaces
47. *P.-T. Liu and J. G. Sutinen,* Control Theory in Mathematical Economics: Proceedings of the Third Kingston Conference, Part B
48. *C. Byrnes,* Partial Differential Equations and Geometry
49. *G. Klambauer,* Problems and Propositions in Analysis
50. *J. Knopfmacher,* Analytic Arithmetic of Algebraic Function Fields
51. *F. Van Oystaeyen,* Ring Theory: Proceedings of the 1978 Antwerp Conference
52. *B. Kedem,* Binary Time Series
53. *J. Barros-Neto and R. A. Artino,* Hypoelliptic Boundary-Value Problems
54. *R. L. Sternberg, A. J. Kalinowski, and J. S. Papadakis,* Nonlinear Partial Differential Equations in Engineering and Applied Science
55. *B. R. McDonald,* Ring Theory and Algebra III: Proceedings of the Third Oklahoma Conference
56. *J. S. Golan,* Structure Sheaves over a Noncommutative Ring
57. *T. V. Narayana, J. G. Williams, and R. M. Mathsen,* Combinatorics, Representation Theory and Statistical Methods in Groups: YOUNG DAY Proceedings
58. *T. A. Burton,* Modeling and Differential Equations in Biology
59. *K. H. Kim and F. W. Roush,* Introduction to Mathematical Consensus Theory
60. *J. Banas and K. Goebel,* Measures of Noncompactness in Banach Spaces
61. *O. A. Nielsen,* Direct Integral Theory
62. *J. E. Smith, G. O. Kenny, and R. N. Ball,* Ordered Groups: Proceedings of the Boise State Conference
63. *J. Cronin,* Mathematics of Cell Electrophysiology
64. *J. W. Brewer,* Power Series Over Commutative Rings
65. *P. K. Kamthan and M. Gupta,* Sequence Spaces and Series
66. *T. G. McLaughlin,* Regressive Sets and the Theory of Isols
67. *T. L. Herdman, S. M. Rankin, III, and H. W. Stech,* Integral and Functional Differential Equations
68. *R. Draper,* Commutative Algebra: Analytic Methods
69. *W. G. McKay and J. Patera,* Tables of Dimensions, Indices, and Branching Rules for Representations of Simple Lie Algebras
70. *R. L. Devaney and Z. H. Nitecki,* Classical Mechanics and Dynamical Systems
71. *J. Van Geel,* Places and Valuations in Noncommutative Ring Theory
72. *C. Faith,* Injective Modules and Injective Quotient Rings
73. *A. Fiacco,* Mathematical Programming with Data Perturbations I
74. *P. Schultz, C. Praeger, and R. Sullivan,* Algebraic Structures and Applications: Proceedings of the First Western Australian Conference on Algebra
75. *L. Bican, T. Kepka, and P. Nemec,* Rings, Modules, and Preradicals
76. *D. C. Kay and M. Breen,* Convexity and Related Combinatorial Geometry: Proceedings of the Second University of Oklahoma Conference
77. *P. Fletcher and W. F. Lindgren,* Quasi-Uniform Spaces
78. *C.-C. Yang,* Factorization Theory of Meromorphic Functions
79. *O. Taussky,* Ternary Quadratic Forms and Norms
80. *S. P. Singh and J. H. Burry,* Nonlinear Analysis and Applications
81. *K. B. Hannsgen, T. L. Herdman, H. W. Stech, and R. L. Wheeler,* Volterra and Functional Differential Equations

82. *N. L. Johnson, M. J. Kallaher, and C. T. Long,* Finite Geometries: Proceedings of a Conference in Honor of T. G. Ostrum
83. *G. I. Zapata,* Functional Analysis, Holomorphy, and Approximation Theory
84. *S. Greco and G. Valla,* Commutative Algebra: Proceedings of the Trento Conference
85. *A. Fiacco,* Mathematical Programming with Data Perturbations II

Other Volumes in Preparation

Mathematical Programming with Data Perturbations II

edited by

Anthony V. Fiacco

Department of Operations Research
The George Washington University
School of Engineering and
Applied Science
Washington, D. C.

MARCEL DEKKER, INC. **New York and Basel**

Library of Congress Cataloging in Publication Data
Main entry under title:

Mathematical programming with data perturbations II.

(Lecture notes in pure and applied mathematics ; 85)
"Papers ... presented to the Second Symposium for Mathematical Programming with Data Perturbations, held at the George Washington University in May 1980"--P. iii.
Includes index.
1. Programming (Mathematics)--Addresses, essays, lectures. 2. Perturbation (Mathematics)--Addresses, essays, lectures. I. Fiacco, Anthony V. II. Symposium on Mathematical Programming with Data Perturbations (2nd : 1980 : George Washington University) III. Series.
QA402.5.M356 1983 519.7 82-22055
ISBN 0-8247-1789-9

MARCEL DEKKER, INC.
270 Madison Avenue, New York, New York 10016

Current printing (last digit):
10 9 8 7 6 5 4 3 2 1

PRINTED IN THE UNITED STATES OF AMERICA

PREFACE

As indicated in *Mathematical Programming with Data Perturbations I* (Volume 73 of this series), our motivation is the unification of important concepts and results in mathematical programming sensitivity and stability analysis methodology. This requires the collection of critical masses of recent theoretical and practical results. We initiated this process with the last volume and continue it with the present.

The papers in this volume were selected from those presented to the Second Symposium for Mathematical Programming with Data Perturbations, held at the George Washington University in May 1980. Each paper was carefully refereed according to the usual standards of professional journals.

The reader will find a wide range of important results and applications in this collection of papers, attesting both to the level of mathematical sophistication that has been attained in the theory and the gradual introduction of rudimentary techniques in diverse applications. Widespread and routine applications of incisive results in NLP sensitivity and stability analysis methodology remain to be developed, but at least we see some efforts and these will hopefully encourage others. Important recent theoretical results are presented, including an extension of constant rank and implicit function theorems, continuity and stability bounds results for general classes of infinite dimensional problems, differential stability bounds, and the interrelationship between optimal value conditions and shadow prices for stable and unstable programs. Applications include some results on the interplay between algorithmic and sensitivity information for spatial equilibrium problems, an application of sensitivity analysis results for convergence acceleration in an algorithm for geometric (signomial) programming, and an application involving the manipulation of parameters in least-squares and implicit model equations for significant algorithmic simplifications, and including formulas for sensitivity analysis calculations. Many new directions of research are indicated, both in the theory and applications of sensitivity and stability methodology.

We expect this area to continue its rapid recent development in the near future. As the now existing quite respectable body of theory gradually unifies, key concepts will become more clearly identified and more familiar to researchers, practitioners, and students. This should encourage more research directed to the development of simpler proofs and interpretations of abstract results and an increasing emphasis on applications to problems with special structure. Ample motivation and precedence for this in NLP is the highly developed methodology for LP sensitivity, postoptimal analysis, and parametric programming.

It is a pleasure to acknowledge the many sources of assistance that made this volume possible: the George Washington University, and its School of Engineering and Applied Science, its Department of Operations Research and Institute for Management Science and Engineering for sponsoring and essentially subsidizing the symposium on which this volume is based; the Office of Naval Research, the Army Research Office, and the National Science Foundation for partial support of the effort required to complete this volume; the referees for careful reviews that resulted in significant improvements; my doctoral dissertation student, Jerzy Kyparisis, for assisting me in proofreading the manuscript; Robin Meader, for her usual excellent technical typing and editorial assistance (for both this volume and Volume 73); and last, but most importantly, the participants of the symposium and particularly the contributors to this volume.

Anthony V. Fiacco

CONTENTS

Preface iii

Contributors vi

CHAPTER

1. Theorem of Constant Rank for Lipschitzian Maps 1
A. Auslender

2. Lipschitzian Perturbations of Infinite Optimization Problems 7
Walter Alt

3. On the Continuity of the Optimum Set in Parametric Semiinfinite Programming 23
Bruno Brosowski

4. Optimality Conditions and Shadow Prices 49
Henry Wolkowicz

5. Optimal Value Continuity and Differential Stability Bounds under the Mangasarian-Fromovitz Constraint Qualification 65
Anthony V. Fiacco

6. Iteration and Sensitivity for a Nonlinear Spatial Equilibrium Problem 91
Caulton L. Irwin and Chin W. Yang

7. A Sensitivity Analysis Approach to Iteration Skipping in the Harmonic Mean Algorithm 109
John J. Dinkel, Gary A. Kochenberger, and Danny S. Wong

8. Least Squares Optimization with Implicit Model Equations 131
Aivars Celmiņš

Index 153

CONTRIBUTORS

WALTER ALT Mathematisches Institut der Universität Bayreuth, Bayreuth, Federal Republic of Germany

A. AUSLENDER Université de Clermont II, Aubiere, France

BRUNO BROSOWSKI Johann Wolfgang Goethe Universität, Frankfurt, Federal Republic of Germany

AIVARS CELMIŅŠ Ballistic Research Laboratory, Aberdeen Proving Ground, Maryland

JOHN J. DINKEL Texas A&M University, College Station, Texas

ANTHONY V. FIACCO The George Washington University, Washington, D.C.

CAULTON L. IRWIN West Virginia University, Morgantown, West Virginia

GARY A. KOCHENBERGER The Pennsylvania State University, University Park, Pennsylvania

HENRY WOLKOWICZ The University of Alberta, Edmonton, Alberta, Canada

DANNY S. WONG Ohio State University, Columbus, Ohio

CHIN W. YANG* West Virginia University, Morgantown, West Virginia

Current affiliation: Clarion State College, Clarion, Pennsylvania

CHAPTER 1 THEOREM OF CONSTANT RANK FOR LIPSCHITZIAN MAPS

A. AUSLENDER / Université de Clermont II, Aubiere, France

ABSTRACT

In this paper the usual theorem of constant rank and the general implicit function theorem are extended to Lipschitzian maps.

I. PRELIMINARIES

In sensitivity analysis with continuously differentiable maps it is well known that the implicit theorem plays a fundamental role. We are interested here in extending this theorem for Lipschitzian maps with an assumption of constant rank.

In differentiable geometry there are two important theorems of constant rank: the usual theorem of constant rank (Dieudonné [4, Theorem 10.3.1]) and the general implicit function theorem (Auslander and MacKenzie [1, 2.2]). These two theorems are given for continuously differentiable maps. In Section 2 we shall extend these theorems for locally Lipschitz functions. In this section we shall now recall some fundamental definitions about Lipschitzian functions.

Let $f : \mathbb{R}^m \to \mathbb{R}^n$ satisfy a Lipschitz condition in a neighborhood of a point x in $\mathbb{R}^m$. Then it is a consequence of Rademacher's theorem that f is almost everywhere differentiable near x.

The usual $m \times n$ Jacobian matrix of partial derivatives at x when it exists is denoted by $J(f;x)$. F. H. Clarke defines in [2] the generalized Jacobian matrix of f at x denoted by $\mathcal{J}(f;x)$ as the convex hull of all matrices M of the form:

$$M = \lim_{i\to\infty} J(f;x_i)$$

where x_i converges to x, and f is differentiable at x_i for each i. We shall say that $\mathcal{J}(f;x)$ is of rank r if every matrix $M \in \mathcal{J}(f;x)$ is of rank r.

Let us now consider a function $\gamma\colon \mathbb{R}^p \times \mathbb{R}^q \to \mathbb{R}^q$ defined on an open set $\mathcal{O}$ in $\mathbb{R}^p \times \mathbb{R}^q$, Lipschitzian in a neighborhood of $(x_0,y_0) \in \mathcal{O}$. Following Hiriart-Urruty [5] we denote $\hat{\mathcal{J}}_y(\gamma;\ (x_0,y_0))$ the convex hull of all matrices M of the form

$$M = \lim_{i\to\infty} J_y(\gamma;\ (x_i,y_i))$$

where $\{(x_i,y_i)\}$ converges to (x_0,y_0), γ is differentiable at (x_i,y_i) for each i, and $J_y(\gamma;\ (x_i,y_i))$ denotes the usual partial Jacobian matrix. For the following we shall need the next implicit function theorem proved by Hiriart-Urruty in [5].

THEOREM 1 Let $(x_0,y_0) \in \mathcal{O}$ be such that $\gamma(x_0,y_0) = 0$. We suppose that $\hat{\mathcal{J}}_y(\gamma;\ (x_0,y_0))$ is of rank q. Then there exists an open neighborhood $V_{(x_0,y_0)}$ of (x_0,y_0) contained in $\mathcal{O}$, an open neighborhood W_{x_0} of x_0, and a function $\sigma\colon W_{x_0} \to \mathbb{R}^q$ Lipschitz in a neighborhood of x_0 such that the relation

(a) $(x,y) \in V_{(x_0,y_0)}$ and $\gamma(x,y) = 0$ is equivalent to the relation:

(b) $x \in W_{x_0}$ and $y = \sigma(x)$.

II. THEOREMS OF CONSTANT RANK

A. A Preliminary Lemma

LEMMA 1 Let $h\colon \mathbb{R}^{n+m} \to \mathbb{R}^q$ be locally Lipschitz, V a nonempty open set in $\mathbb{R}^n$, and W a nonempty open and connected set in $\mathbb{R}^m$. We assume that the partial Jacobian matrix $J_y(h;\cdot)$ is almost everywhere equal to zero in $V \times W$. Then for each $x \in V$ the mapping $y \to h(x,y)$ is constant on W.

PROOF. 1. Let $y \in W$. Since W is open there exists $\rho_y > 0$ such that the open ball $B_y = B(y,\rho_y)$ centered in y with radius ρ_y is in W. Then let

$u \in B_y$, $u \neq y$; set $v = (u - y)/||u - y||$, then $u = y + \delta v$ with $\delta = ||u - y||$. Now there exists $r > 0$ such that

$$y + w + tv \in B_y \qquad \forall\, t \in [0,\delta],\ w \in \bar{B}(0,r).$$

Let a subset S of $\mathbb{R}^m$ have measure 0 and consider the set of lines in $\mathbb{R}^m$ parallel to a given vector. As was pointed out by Clarke [3, proof of Proposition 1.4], it is a consequence of Fubini's theorem that almost all of these lines meet S in a set of one-dimensional measure equal to zero. If we apply this fact where S is the set of points at which $J_y(h;\cdot)$ fails to exist and to be equal to zero, we deduce that for almost all $x \in V$, $w \in \bar{B}(0,r)$ the function $t \to h(x,y + w + tv)$ is differentiable a.e., its derivative being $J_y(h;\ (x,y + w + tv)) \cdot v = 0$ a.e. We have then:

$$h(x,y + w + \delta v) - h(x,y + w) = \int_0^\delta J_y(h;\ (x,y + w + tv)) \cdot v\, dt = 0$$

Since h is continuous, this then must hold for all $(x,w) \in V \times \bar{B}(0,r)$ without exception, and we obtain

$$h(x,u) = h(x,y).$$

2. Now let $\bar{y}$ be a fixed point in W and denote $\Lambda(x,\bar{y}) = \{y \in W\colon h(x,y) = h(x,\bar{y})\}$. Since h is continuous, $\Lambda(x,\bar{y})$ is closed for the induced topology on W. From part 1, if $y \in \Lambda(x,\bar{y})$, there exists an open ball B_y in $\Lambda(x,\bar{y})$. Since W is a connected set and since $\Lambda(x,\bar{y})$ is nonempty, then $\Lambda(x,\bar{y})$ is equal to W.

B. Theorems of Constant Rank

Let $\mathcal{O}$ be a nonempty open subset in $\mathbb{R}^n$, X a Lipschitzian function from $\mathcal{O}$ into $\mathbb{R}^s$ with components $X_1, X_2, \ldots, X_s$. Let ℓ be a positive integer such that $\ell \leqq \min(n,s)$. For each $u \in \mathcal{O}$ and each $x \in \mathbb{R}^s$, set:

$$v = (u_1, u_2, \ldots, u_\ell), \qquad w = (u_{\ell+1}, \ldots, u_n)$$

$$y = (x_1, x_2, \ldots, x_\ell), \qquad z = (x_{\ell+1}, \ldots, x_s)$$

$$Y(v,w) = (X_1(u), \ldots, X_\ell(u)), \qquad Z(v,w) = (X_{\ell+1}(u), \ldots, X_s(u)).$$

Now let $\bar{u} \in \mathcal{O}$ and $\bar{v} = (\bar{u}_1, \ldots, \bar{u}_\ell)$, $\bar{w} = (\bar{u}_{\ell+1}, \ldots, \bar{u}_n)$.

THEOREM 2 Let E be a subset of $\mathcal{O}$ of measure 0. We suppose that for each $u \in \mathcal{O}\setminus E$, $\mathcal{J}(X;u)$ is of rank ℓ and that $\hat{\mathcal{J}}_v(Y;\ (\bar{v},\bar{w}))$ is of rank ℓ. Then:

(A) there exists an open set $\mathcal{U}$ contained in $\mathcal{O}$ containing $\bar{u}$ and a Lipschitzian function g defined on an open subset $\mathcal{Y}$ containing $Y(\bar{u})$ of R^{ℓ} into $R^{s-\ell}$ such that $X(\mathcal{U})$ can be considered as the graph of g; that is,

$x \in X(\mathcal{U}) \iff y \in \mathcal{Y},\ z = g(y)$;

(B) there exists a neighborhood W of $\bar{w}$, a neighborhood V of $\bar{v}$, and Lipschitzian function σ: $W \to V$ such that

1. the relation $(v,w) \in V \times W$ and $Y(v,w) = Y(\bar{u})$ is equivalent to the relation $w \in W$ and $v = \sigma(w)$

2. $Z(\sigma(w),w) = Z(\bar{u}) \qquad \forall\, w \in W$.

PROOF.

(A) 1. Let $\bar{x} = X(\bar{u})$, $\bar{y} = Y(\bar{u})$, $\bar{z} = Z(\bar{u})$. From Theorem 1 there exist open connected sets V, W, $\mathcal{Y}$ containing, respectively, $\bar{v}$, $\bar{w}$, $\bar{y}$ and a Lipschitzian function θ: $\mathcal{Y} \times W \to V$ such that the relation

(a) $v \in V,\ w \in W,\ y \in \mathcal{Y}$ and $Y(v,w) = y$

is equivalent to the relation

(b) $y \in \mathcal{Y},\ w \in W$ and $v = \theta(y,w)$.

Let $\mathcal{U} = (V \times W) \cap Y^{-1}(\mathcal{Y})$ and define on $\mathcal{Y} \times W$ the function

$$h(y,w) = Z(\theta(y,w),w)$$

Theorem 2 will be proved if we show that

$$h(y,w) = h(y,w_0) \qquad \forall\, y \in \mathcal{Y},\ \forall\, w \in W \tag{1}$$

w_0 being an arbitrary point of W. Indeed, in this case we shall set $g(y) = h(y,w_0)$, and for each $x = (y,z) \in X(\mathcal{U})$ from the equivalence between (a) and (b), this will imply that $z = g(y)$. Equality (1) will be proved if by using the fundamental lemma we can obtain a set S of measure 0 in $\mathcal{Y} \times W$ such that $J_w(h;\cdot)$ is equal to zero in the complement of S in $\mathcal{Y} \times W$.

2. Construction of S. Let $D = \{(y,w) \in \mathcal{Y} \times W$: θ is nondifferentiable in $(y,w)\}$. By Rademacher's theorem D is a set of measure 0. Let $F = \{(v,w) \in \mathcal{U}$: X is nondifferentiable in $(v,w)\}$; then F is also a set of measure 0. Let $\bar{\theta}$ be the mapping defined on $\mathcal{Y} \times W$ by

$$\bar{\theta}(y,w) = (\theta(y,w),w)$$

and $\bar{\theta}^{-1}$ be the mapping defined on $\mathcal{U}$ by

$\overline{\theta}^{-1}(v,w) = (Y(v,w),w).$

Then $\overline{\theta}^{-1}$ is a Lipschitzian map with values in $\mathbb{R}^n$; this implies that the image by $\overline{\theta}^{-1}$ of a set of measure 0 is again a set of measure 0. From this it follows that the sets $\overline{\theta}^{-1}(F)$ and $\overline{\theta}^{-1}(E')$, where $E' = E \cap \mathcal{U}$ ($\mathcal{U}$) are of measure 0. Now we can define S as follows:

$$S = D \cup \overline{\theta}^{-1}(F) \cup \overline{\theta}^{-1}(E')$$

3. Now let $(y,w) \in \mathcal{Y} \times W$, $(y,w) \notin S$. Part (A) will be proved if we can show that $J_w(h; (y,w)) = 0$. Let $v = \theta(y,w)$.

(α) Since $(y,w) \notin D$, θ is differentiable in (y,w).

(β) Since $(y,w) \notin \overline{\theta}^{-1}(F)$, Y and Z are differentiable in (v,w).

(γ) Now let Y_i, Z_j, θ_k, h_j be the components of Y, Z, θ, h.

By the chain rule, h_i is differentiable in (y,w) and we have:

$$\frac{\partial h_i}{\partial w_j}(y,w) = \sum_{k=1}^{\ell} \frac{\partial Z_i}{\partial v_k}(v,w)\frac{\partial \theta_k}{\partial w_j}(y,w) + \frac{\partial Z_i}{\partial w_j}(v,w). \tag{2}$$

Now using the same argument as in Lemma 3 of [2], we can show that there exists a neighborhood $\mathcal{U}'$ of $\bar{u}$ such that $\hat{J}_v(Y; (v,w))$ is of rank ℓ, for each $(v,w) \in \mathcal{U}'$. Without loss of generality we can suppose that $\mathcal{U}' = \mathcal{U}$. Then since $(y,w) \notin \overline{\theta}^{-1}(E')$, (v,w) does not belong to E and by the rank assumption there exist functions $a_j^i: V \times W \to \mathbb{R}$, $i \in \langle 1, s-\ell\rangle$, $j \in \langle 1,\ell\rangle$ such that:

$$\frac{\partial Z_i}{\partial v_j}(v,w) = \sum_{m=1}^{\ell} a_m^i(v,w)\frac{\partial Y_m}{\partial v_j}(v,w) \qquad \forall\, j \in \langle 1,\ell\rangle \tag{3}$$

$$\frac{\partial Z_i}{\partial w_j}(v,w) = \sum_{m=1}^{\ell} a_m^i(v,w)\frac{\partial Y_m}{\partial w_j}(v,w) \qquad \forall\, j \in \langle 1, n-\ell\rangle \tag{4}$$

Combining formulas (2), (3), (4) we obtain

$$\begin{aligned}\frac{\partial h_i}{\partial w_j}(y,w) &= \sum_{k=1}^{\ell}\sum_{m=1}^{\ell} a_m^i(v,w)\frac{\partial Y_m}{\partial v_k}(v,w)\frac{\partial \theta_k}{\partial w_j}(y,w) \\ &\quad + \sum_{m=1}^{\ell} a_m^i(v,w)\frac{\partial Y_m}{\partial w_j}(v,w) \\ &= \sum_{m=1}^{\ell} a_m^i(v,w)\left[\sum_{k=1}^{\ell}\frac{\partial Y_m}{\partial v_k}(v,w)\cdot\frac{\partial \theta_k}{\partial w_j}(y,w) + \frac{\partial Y_m}{\partial w_j}(v,w)\right].\end{aligned}$$

But from the equivalence between (a) and (b) we have

$$Y_m(\theta(y,w),w) - y = 0 \qquad \forall\, y \in \mathcal{Y},\ \forall\, w \in W.$$

Using the chain rule, if we differentiate this expression we obtain

$$\sum_{k=1}^{\ell} \frac{\partial Y_m}{\partial v_k}(v,w)\, \frac{\partial \theta_k}{\partial w_j}(y,w) + \frac{\partial Y_m}{\partial w_j}(v,w) = 0$$

so that

$$\frac{\partial h_i}{\partial w_j}(y,w) = 0 \qquad \forall\, i.$$

(B) Set $\bar{y} = Y(\bar{u})$, $\bar{z} = Z(\bar{u})$; if we define $\sigma(\cdot)$ by

$$\sigma(w) = \theta(\bar{y},w)$$

then from the equivalence between (a) and (b) we obtain the equivalence between the relation $v \in V$, $w \in W$, $Y(v,w) = Y(\bar{u})$, and the relation $v = \sigma(w)$, $w \in W$. From relation (1) applied with $y = \bar{y}$ we obtain, since $\sigma(\bar{w}) = \bar{v}$,

$$Z(\sigma(w),w) = Z(\sigma(\bar{w}),\bar{w}) = Z(\bar{u}) \qquad \forall\, w \in W.$$

REMARK. Part (A) of Theorem 2 is an extension of the constant rank theorem of Dieudonné [4, Theorem 10.3.1] to Lipschitzian maps, whereas part (B) is an extension of the generalized implicit function theorem [1, Section 2.2].

REFERENCES

1. L. Auslender and R. E. MacKenzie. *Introduction to Differentiable Manifolds.* McGraw-Hill, Hightstown, New Jersey, 1963.

2. F. H. Clarke. On the inverse function theorem, *Pacific Journal of Mathematics*, Vol. 64, pp. 97-102 (1976).

3. F. H. Clarke. Necessary conditions for nonsmooth problems in optimal control and the calculus of variations. Ph.D. dissertation, University of Washington, 1973.

4. J. Dieudonné. *Foundations of Modern Analysis.* Academic Press, New York, 1960.

5. J.-B. Hiriart-Urruty. Tangent cones, generalized gradients and mathematical programming in Banach spaces, *Mathematics of Operations Research*, Vol. 4, pp. 79-97 (1979).

CHAPTER 2 LIPSCHITZIAN PERTURBATIONS OF INFINITE OPTIMIZATION PROBLEMS

WALTER ALT / Mathematisches Institut der Universität Bayreuth, Bayreuth, Federal Republic of Germany

ABSTRACT

This paper is concerned with infinite-dimensional nonlinear optimization problems in which the objective function and the constraints are subjected to Lipschitzian perturbations. In particular, continuity properties of the extremal value function are studied and bounds for the distance between local minimizers of perturbed problems and the original problem are derived.

I. INTRODUCTION AND PRELIMINARIES

This paper deals with continuity properties of the extremal value function and the stability of solutions of infinite nonlinear optimization problems when the data of the problem are subjected to perturbations. The problem of interest is given by real Banach spaces X and Y, a metric space (W,δ), closed convex subsets C and K of X and Y, respectively, and mappings $f\colon A \times W \to \mathbb{R}$, $g\colon A \times W \to Y$, where A is an open set containing C. For each fixed $w \in W$ we consider the following optimization problem:

$$\text{Minimize } f(x,w) \qquad \text{subject to } x \in \Sigma(w) \tag{P_w}$$

Here $\Sigma(w)$ is the set of feasible points for (P_w), i.e.,

$$\Sigma(w) = \{x \in C \cap A \mid g(x,w) \in K\}$$

In this way a family $(P_w)_{w \in W}$ of infinite optimization problems is defined with perturbation parameter w and parameter set W.

For some specific parameter $w_0 \in W$ the problem (P_{w_0}) is looked upon as the original unperturbed problem. We assume that (P_{w_0}) has a local solution x_0 and define the set $\Sigma_r(w)$ by

$$\Sigma_r(w) = \Sigma(w) \cap B(x_0,r)$$

where $r \geqq 0$ and $B(x_0,r)$ is the closed ball with radius r around x_0. The extremal value function $\mu_r: W \to \overline{\mathbb{R}}$ is defined by

$$\mu_r(w) = \inf \{f(x,w) \mid x \in \Sigma_r(w)\}$$

here $\overline{\mathbb{R}} = \mathbb{R} \cup \{\infty\} \cup \{-\infty\}$, $\inf \emptyset = \infty$, and $\inf A = -\infty$, if A is not bounded from below.

Continuity properties of the extremal value function are of great interest in many applications. Our first aim is to give sufficient conditions for the Lipschitz continuity of μ_r at w_0. Similar problems were investigated by many authors and some of their results are closely related to ours. The papers by Fiacco [2], Gauvin [3], Robinson [11,12], and Sargent [13] are concerned with finite dimensional problems; Levitin [5] considers infinite dimensional problems with an operator equality and finitely many inequality constraints. We deal with more general problems by allowing constraints defined by arbitrary convex sets.

Our second aim is to give bounds for the distance between local minimizers of perturbed problems and the local minimizer x_0 of the unperturbed problem. This is done under an assumption, which is closely related to sufficient optimality conditions. Similar results for second order sufficient optimality conditions can be found in the papers of Levitin [5] and Robinson [12].

II. REGULARITY AND STABILITY OF DIFFERENTIABLE NONLINEAR SYSTEMS

In this section we state a stability theorem for differentiable nonlinear systems, which we shall use in the sequel in the proofs of our main results.

First we introduce some notions. Let (E,δ) be a metric space. We use the abbreviation *int* for (topological) interior; for $x \in E$ and $A \subset E$ let

$$d[x,A] = \inf \{\delta(x,a) \mid a \in A\} \qquad (= \infty \text{ if } A = \emptyset)$$

In view of the following definition we recall that problem (P_{w_0}) is looked upon as the original unperturbed problem.

DEFINITION 1 Let $x_0 \in \Sigma(w_0)$, i.e., $x_0 \in C \cap A$ and $g(x_0,w_0) \in K$. Then x_0 is called *regular* if

(2.1) For each $w \in W$ the mapping $g(\cdot,w)$ is Gateaux-differentiable in a neighborhood of x_0; the Gateaux-differential is denoted by g_x.

(2.2) The mappings g and g_x are continuous at (x_0,w_0).

(2.3) $0 \in \text{int}\, \{g(x_0,w_0) + g_x(x_0,w_0)(C - x_0) - K\}$.

REMARK 1.

(a) Assumption (2.2) implies Fréchet-differentiability of $g(\cdot,w_0)$ at (x_0,w_0).

(b) In (2.3) it suffices to require that the origin be an internal point of the set in question; compare in this connection Robinson [9,10].

(c) In some special cases it can be shown that (2.3) is equivalent to well known necessary optimality conditions such as the Slater condition and the Mangasarian-Fromovitz condition.

We can now formulate the mentioned result on stability of differentiable nonlinear systems.

THEOREM 1 Let $x_0 \in \Sigma(w_0)$ be a regular point. Define a multi-function $G: X \times W \to Y$ by

$$G(x,w) = \begin{cases} g(x,w) - K & x \in C \\ \emptyset & x \notin C \end{cases}$$

Then for any $\varepsilon > 0$ there exist neighborhoods U_ε of x_0 and V_ε of w_0 such that for each $w \in V_\varepsilon$ the set $\Sigma(w)$ is nonempty and for each such w and each $x \in U_\varepsilon$ we have

$$d[x,\Sigma(w)] \leqq (L + \varepsilon)d[0,G(x,w)] \tag{2.4}$$

with some constant $L > 0$, which can be chosen independently of ε.

This result is a generalized version of Theorem 1 in Robinson [9]. A proof can be found in Maguregui [7]. The result can also be derived from Theorem II.2.8 in Alt [1] (see the appendix).

For a real number $r > 0$ we define

$$C_r = C \cap B(x_0,r)$$

By Theorem 2.1 in Zowe [14] the regularity condition (2.3) is equivalent to the following statement

$$0 \in \text{int}\,\{g(x_0,w_0) + g_x(x_0,w_0)(C_r - x_0) - K\} \tag{2.5}$$

As an immediate consequence we get the following corollary to Theorem 1.

COROLLARY 1 Let $r > 0$ be a real number and let $x_0 \in \Sigma_r(w_0)$ be a regular point. Define a multifunction $G_r : X \times W \to Y$ by

$$G_r(x,w) = \begin{cases} g(x,w) - K & x \in C_r \\ \emptyset & x \notin C_r \end{cases}$$

Then for any $\varepsilon > 0$ there exist neighborhoods U_ε of x_0 and V_ε of w_0 such that for each $w \in V_\varepsilon$ the set $\Sigma_r(w)$ is nonempty and for each such w and each $x \in U_\varepsilon$ we have

$$d[x,\Sigma_r(w)] \leqq (L + \varepsilon)d[0,G_r(x,w)] \tag{2.6}$$

with some constant $L > 0$ which can be chosen independently of ε.

III. CONTINUITY PROPERTIES OF THE EXTREMAL VALUE FUNCTION

In the following we shall study the local behavior of the extremal value function near w_0. The first result is a statement on upper semicontinuity.

THEOREM 2 Let x_0 be a local minimizer of problem (P_{w_0}), i.e., there is some $r > 0$ such that

$$f(x_0,w_0) \leqq f(x,w_0) \qquad \text{for all } x \in \Sigma_r(w_0)$$

Suppose x_0 is regular and f is continuous at (x_0,w_0). Then the extremal value function μ_r is upper semicontinuous at w_0, i.e.,

$$\limsup_{w \to w_0} \mu_r(w) \leqq \mu_r(w_0) = f(x_0,w_0)$$

PROOF. Since x_0 is regular we can apply Corollary 1 with $\varepsilon = 1$. Hence, there exists a neighborhood V of w_0 and a constant $L > 0$ such that for each $w \in V$ the set $\Sigma_r(w)$ is nonempty and (2.6) implies

$$d[x_0,\Sigma_r(w)] \leqq (L + 1)||g(x_0,w) - g(x_0,w_0)|| \tag{3.1}$$

Now suppose $\{w_n\} \subset W$ is a sequence with

$$\lim_{n\to\infty} w_n = w_0 \text{ and}$$

$$\lim_{n\to\infty} \mu_r(w_n) = \limsup_{w\to w_0} \mu_r(w)$$

Then for sufficiently large n we have $\Sigma_r(w_n) \neq \emptyset$ and by (3.1) there exist points $x_n \in \Sigma_r(w_n)$ such that

$$||x_n - x_0|| \leqq 2(L + 1)||g(x_0,w_n) - g(x_0,w_0)|| \tag{3.2}$$

By the continuity of g at (x_0,w_0) we conclude from (3.2) that $\lim_{n\to\infty} x_n = x_0$ and therefore $\lim_{n\to\infty} f(x_n,w_n) = f(x_0,w_0)$ because of the continuity of f at (x_0,w_0). From the inequality $f(x_n,w_n) \geqq \mu_r(w_n)$ we finally obtain

$$f(x_0,w_0) = \mu_r(w_0) = \lim_{n\to\infty} f(x_n,w_n) \geqq \lim_{n\to\infty} \mu_r(w_n) = \limsup_{w\to w_0} \mu_r(w)$$

In the above theorem we only required continuity of the mappings f and g at (x_0,w_0). If we impose more restrictive assumptions on these mappings, our result can be improved. The next theorem shows that if f and g are Lipschitz, this property is carried over to the extremal value function.

THEOREM 3 Let x_0 be a local minimizer of (P_{w_0}). Suppose x_0 is regular and there is a neighborhood $\tilde{U} = U_x \times U_w$ of (x_0,w_0) such that

$$|f(x,w) - f(\bar{x},w_o)| \leqq L_f(||x-\bar{x}|| + \delta(w,w_o)) \tag{3.3}$$

for all (x,w) $\tilde{U}$ and all $\bar{x}$ U_x with some constant $L_f < 0$, i.e., f is Lipschitz in $\tilde{U}$

$$||g(x,w) - g(x,w_0)|| \leqq L_g\delta(w,w_0) \tag{3.4}$$

for all $(x,w) \in \tilde{U}$ with some constant $L_g > 0$, i.e., for each $x \in U_x$ the mapping $g(x,\cdot)$ is Lipschitz at w_0.

Then there exist a real number $r > 0$ and a neighborhood V of w_0 with

$$\mu_r(V) \subset \mathbb{R} \tag{3.5}$$

i.e., μ_r is finite on V and

$$|\mu_r(w) - \mu_r(w_0)| \leqq L_r\delta(w,w_0) \tag{3.6}$$

for all $w \in V$ with some constant $L_r > 0$, i.e., μ_r is Lipschitz at w_0.

PROOF. Since x_0 is a local minimizer of (P_{w_0}), we have

$$f(x,w_0) \leqq f(x_0,w_0) \qquad \text{for all } x \in \Sigma_s(w_0) \tag{*}$$

with a real number $s > 0$. We choose $r > 0$ and a neighborhood V of w_0 such that

(a) $r \leqq s$;
(b) $B(x_0,r) \subset U_1$, $V \subset V_1$ with the sets U_1, V_1, which are determined by Corollary 1;
(c) $B(x_0,r) \subset U_x$;
(d) $2(L + 1)L_g\ \delta(w,w_0) < r$, for all $w \in V$, where L is the constant from Corollary 1.

Then for arbitrary $w \in V$ we have, by Corollary 1 and (3.4),

$$\begin{aligned} d[x_0,\Sigma_s(w)] &\leqq (L + 1)||g(x_0,w) - g(x_0,w_0)|| \\ &\leqq (L + 1)L_g\ \delta(w,w_0) \end{aligned} \tag{3.7}$$

This implies the existence of a point $x_w \in \Sigma_s(w)$ with

$$||x_w - x_0|| \leqq 2(L + 1)L_g\ \delta(w,w_0) \tag{3.8}$$

By (d) it follows that $x_w \in \Sigma_r(w)$, and therefore

$$f(x_w,w) \geqq \mu_r(w) \tag{3.9}$$

Let

$$L_r = L_f(2(L + 1)L_g + 1) \tag{3.10}$$

From (3.8) and (3.3) we have

$$|f(x_w,w) - f(x_0,w_0)| \leqq L_r\ \delta(w,w_0)$$

Combining this with (3.9) we get

$$\begin{aligned} \mu_r(w) &\leqq f(x_w,w) \leqq f(x_0,w_0) + L_r\ \delta(w,w_0) \\ &= \mu_r(w_0) + L_r\ \delta(w,w_0) \qquad \text{for all } w \in V \end{aligned} \tag{3.11}$$

Next we prove the inequality

$$\mu_r(w_0) \leqq \mu_r(w) + L_r\ \delta(w,w_0) \qquad \text{for all } w \in V \tag{3.12}$$

Let $w \in V$ be arbitrary and $x \in \Sigma_r(w)$. Then $g(x,w) \in K$, and therefore $g(x,w_0) - g(x,w) \in G_s(x,w_0)$. Applying Corollary 1 it follows that

$$
\begin{aligned}
d[x,\Sigma_s(w_0)] &\leqq (L + 1)d[0,G_s(x,w_0)] \\
&\leqq (L + 1)||g(x,w) - g(x,w_0)|| \\
&\leqq (L + 1)L_g \ \delta(w,w_0)
\end{aligned}
$$

Hence there exists a point $\bar{x} \in \Sigma_s(w_0)$ with the property

$$||x - \bar{x}|| \leqq 2(L + 1)L_g \ \delta(w,w_0) \tag{3.13}$$

and

$$f(\bar{x},w_0) \geqq f(x_0,w_0) \tag{3.14}$$

because of (*). In addition, the following inequality is valid by (3.3), (3.10), and (3.13)

$$|f(x,w) - f(\bar{x},w_0)| \leqq L_f(||x - \bar{x}|| + \delta(w,w_0)) \leqq L_r \ \delta(w,w_0) \tag{3.15}$$

From this and (3.14) one easily obtains

$$\mu_r(w_0) = f(x_0,w_0) \leqq f(\bar{x},w_0) \leqq f(x,w) + L_r \ \delta(w,w_0)$$

which proves (3.12). Finally, combining (3.11) and (3.12) we get the desired result.

In conclusion in this section we discuss some related results. Levitin [5] considers infinite dimensional problems of the form (P_w), with a set K of the special form $\{0\} \times \mathbb{R}^m_-$ where 0 is the origin of a Banach space Z and $\mathbb{R}^m_-$ is the negative orthant in $\mathbb{R}^m$. In this situation his regularity assumption is equivalent to the statement (2.3), as we mentioned in Remark 1 above. The Lipschitz conditions imposed on the objective and constraint functions are somewhat different from those required in Theorem 3. In his papers [5,6], Levitin also considers differential properties of the marginal function. Using Theorem 1 or Theorem II.2.8 in Alt [1], these results can likewise be extended for more general problems. This was done by Lempio and Maurer [4]. Gauvin [3] studies finite dimensional problems. His regularity assumption is the well known Mangasarian-Fromovitz condition, which again appears to be equivalent to (2.3) in the underlying situation. Using suitable compactness criteria, Gauvin can also prove lower semicontinuity and hence continuity of the marginal function. The result of [3], Theorem 5.1, is somewhat sharper than that given in our Theorem 3. However, the proof in [3] depends substantially on compactness arguments not present in infinite dimensions; therefore, Gauvin's proof is quite different from that presented here for Theorem 3.

IV. STABILITY OF SOLUTIONS OF PERTURBED OPTIMIZATION PROBLEMS

We continue our analysis with the study of the behavior of local solutions when the data of the optimization problem are subjected to small perturbations. Our aim is to develop sufficient conditions for the stability of local minimizers. Stability here means that the rate of change in the solutions is bounded in terms of the amount of the perturbation. In the last section we showed that in the presence of a regularity condition the extremal value function behaves well. Nevertheless, the solutions of the perturbed problems can behave badly. This is demonstrated by the following example.

EXAMPLE 1. Let $X = Y = \mathbb{R}^2$, $W = \mathbb{R}$, $C = \{(x_1,x_2) \mid x_1 \leqq 1,\ x_2 \leqq 1\}$, $K = \{(x_1,x_2) \mid x_1 \geqq 0,\ x_2 \geqq 0\}$, $f(x_1,x_2,w) = -x_1 - wx_2$, and $g(x_1,x_2,w) = (x_1,x_2)$. For $w_0 = 0$, the point $x_0 = (1,1/2)$ is a (local) solution of

$$\text{Minimize } f(x,w_0) \qquad \text{subject to } x \in C \text{ and } g(x,w_0) \in K \tag{P_{w_0}}$$

Now let $r \in (0,1/2]$ be arbitrary. We provide X with the norm $||(x_1,x_2)|| = \max\{|x_1|,|x_2|\}$. Then for $w \neq 0$, the perturbed problem

$$\text{Minimize } f(x,w) \qquad \text{subject to } ||x - x_0|| \leqq r,$$

$$x \in C, \text{ and } g(x,w) \in K \tag{P_w}$$

has the unique solution

$$x_w = \begin{cases} \left(1, \dfrac{1}{2} + r\right) & \text{if } w > 0 \\ \left(1, \dfrac{1}{2} - r\right) & \text{if } w < 0 \end{cases}$$

It is easily seen that x_0 is regular and the extremal value function μ_r is given by

$$\mu_r(w) = \begin{cases} -1 - w\left(\dfrac{1}{2} + r\right) & w > 0 \\ -1 - w\left(\dfrac{1}{2} - r\right) & w < 0 \end{cases}$$

This shows that μ_r is Lipschitz at w_0 but the distance $||x_w - x_0||$ is not bounded by the amount of the perturbation.

The reason for the behavior of the optimal solutions in the above example is that the unperturbed problem does not depend on the second variable x_2. Therefore all points of the line segment

$\{(x_1,x_2) \mid x_1 = 1, x_2 \in [0,1]\}$ are optimal. On the other hand, the perturbed problems depend on x_2 and have unique solutions because of the local invertibility of the functions $f(\cdot,w)$ for $w \neq 0$. This fact suggests that some kind of local invertibility should be imposed on f in order to assure stability of the optimal solutions.

In the following we shall derive qualitative bounds for the distance $||x_w - x_0||$ of a local minimizer x_w of (P_w) and the local solution x_0 of (P_{w_0}). In the papers of Levitin [5], Robinson [12], and Sargent [13], a second order sufficient optimality condition is used to derive stability results. Our approach is based on a more general abstract condition closely related to sufficient optimality conditions.

DEFINITION 2 We say that condition (B) is fulfilled iff there are reals $\alpha > 0$, $\beta \geqq 1$, $\gamma > 0$ and neighborhoods U of x_0 and V of w_0 such that for each $w \in V$ the following condition is satisfied:

$$f(x,w) \geqq f(x_0,w_0) + \alpha||x - x_0||^\beta - \gamma\, \delta(w,w_0)$$
$$\text{for all } x \in \Sigma(w) \cap U \tag{4.1}$$

Next we present a sufficient condition for (B).

THEOREM 4 Suppose there exist reals $\bar{\alpha} > 0$, $\bar{\beta} \geqq 1$, and a neighborhood $\bar{U}$ of x_0 with

$$f(x,w_0) \geqq f(x_0,w_0) + \bar{\alpha}||x - x_0||^{\bar{\beta}} \qquad \text{for all } x \in \Sigma(w_0) \cap \bar{U} \tag{4.2}$$

If x_0 is regular and assumptions (3.3) and (3.4) are satisfied, then condition (B) is fulfilled.

PROOF. We choose $r,s > 0$ and V such that conditions (*), (a) - (d) are satisfied and $B(x_0,s) \in \bar{U}$. Let $U = \text{int } B(x_0,r)$ and $x \in \Sigma_r(w_0)$, $w \in V$. Then (3.13) implies the existence of a point $\bar{x} \in \Sigma_s(w_0)$ with

$$||x - \bar{x}|| \leqq 2(L + 1)L_g\, \delta(w,w_0) \tag{4.3}$$

and because of (3.15) it follows that

$$f(x,w) \geqq f(\bar{x},w_0) - L_r\, \delta(w,w_0) \tag{4.4}$$

Now from (4.2) we know that

$$f(\bar{x},w_0) \geqq f(x_0,w_0) + \bar{\alpha}||\bar{x} - x_0||^{\bar{\beta}} \tag{4.5}$$

with reals $\bar{\alpha} > 0$ and $\bar{\beta} \geqq 1$. Using the mean value theorem, we get the inequality

$$(||\bar{x} - x_0|| + ||x - \bar{x}||)^{\bar{\beta}} \leqq ||\bar{x} - x_0||^{\bar{\beta}}$$

$$+ \bar{\beta}(||\bar{x} - x_0|| + ||x - \bar{x}||)^{\bar{\beta}-1}||x - \bar{x}||$$

$$\leqq ||\bar{x} - x_0||^{\bar{\beta}} + \bar{\beta}(3s)^{\bar{\beta}-1}||x - \bar{x}||$$

Together with (4.3), this yields

$$||x - x_0||^{\bar{\beta}} \leqq (||\bar{x} - x_0|| + ||x - \bar{x}||)^{\bar{\beta}}$$

$$\leqq ||\bar{x} - x_0||^{\bar{\beta}} + \bar{\beta}(3s)^{\bar{\beta}-1}2(L + 1)L_g\ \delta(w,w_0) \qquad (4.6)$$

Let $\gamma = \bar{\alpha}\bar{\beta}(3r)^{\bar{\beta}-1}2(L + 1)L_g + L_r$, $\alpha = \bar{\alpha}$, and $\beta = \bar{\beta}$. Then it follows from (4.4) and (4.5) that

$$f(x,w) \geqq f(x_0,w_0) + \alpha||\bar{x} - x_0||^{\beta} - L_r\ \delta(w,w_0)$$

and from (4.6) we get

$$f(x,w) \geqq f(x_0,w_0) + \alpha||x - x_0||^{\beta} - \gamma\ \delta(w,w_0)$$

REMARK 2. It is an obvious fact that condition (B) also implies (4.2).

As mentioned previously, condition (B) is closely related to sufficient optimality conditions. To show this we review a first order sufficient condition.

THEOREM 5 (Maurer [8]) Suppose x_0 is regular, $C = X$, and K is a closed convex cone. Let

$$L(x_0) = \{x \mid g(x_0,w_0) + g(x_0,w_0)(x - x_0) \in K\}$$

Suppose further that

$$f'(x_0)(x - x_0) \geqq \alpha||x - x_0|| \qquad \text{for all } x \in L(x_0)$$

Then (4.2) is satisfied with $\bar{\beta} = 1$.

REMARK 3. It was also shown by Maurer [8] that a second order sufficient optimality condition implies (4.2) with $\bar{\beta} = 2$ and hence (B). In a similar way, suitable sufficient optimality conditions of order n result in (4.2) with $\bar{\beta} = n$.

After these preliminaries we can proceed with the formulation of the main result which we announced at the beginning of this section.

THEOREM 6 Let the assumptions of Theorem 3 and condition (B) be satisfied. Then there exist real numbers $r > 0$, $\tilde{d} > 0$, and a neighborhood V of w_0 such that the extremal value function μ_r is Lipschitz at w_0 and for each $w \in V$ the following statements hold:

(i) For any sequence $\{x_n\} \subset \Sigma_r(w)$ with the property $\lim_{n\to\infty} f(x_n,w) = \mu_r(w)$ it follows that $x_n \in \text{int } B(x_0,r)$ for sufficiently large n.

(ii) If there exists a point $x_w \in \Sigma_r(w)$ with $\mu_r(w) = f(x_w,w)$, then $x_w \in \text{int } B(x_0,r)$, i.e., x_w is a local minimizer of (P_w), and

$$||x_w - x_0|| \leqq \tilde{\alpha}\, \delta(w,w_0)^{1/\beta}$$

PROOF. We choose $r > 0$ and neighborhood V of w_0 with the properties (a) - (d) in the proof of Theorem 3, and

(e) condition (B) holds for $x \in B(x_0,r)$, $w \in V$;

(f) $\alpha r^\beta > (L_r + \gamma)\delta(w,w_0)$ for all $w \in V$ with the reals α, β, γ from (B) and the Lipschitz constant L_r of μ_r determined by Theorem 3.

Then μ_r is Lipschitz at w_0 with Lipschitz constant L_r by Theorem 3. Now let $w \in V$ be arbitrary and let $\{x_n\} \subset \Sigma_r(w)$ be a sequence with $\lim_{n\to\infty} f(x_n,w) = \mu_r(w)$. Using assumption (B) and the Lipschitz continuity of μ_r, we get

$$\begin{aligned}\alpha||x_n - x_0||^\beta &\leqq |f(x_n,w) - f(x_0,w_0)| + \gamma\delta(w,w_0)\\ &\leqq |f(x_n,w) - \mu(w)| + |\mu(w) - \mu(w_0)| + \gamma\delta(w,w_0)\\ &\leqq |f(x_n,w) - \mu(w)| + (L_r + \gamma)\delta(w,w_0)\end{aligned}$$

The first term of the right-hand side tends to 0; hence because of assumption (f), part (i) of the theorem is proven.

For a point $x_w \in \Sigma_r(w)$ with $f(x_w,w) = \mu_r(w)$ we conclude by the same consideration

$$\alpha||x_w - x_0||^\beta \leqq (L_r + \gamma)\delta(w,w_0)$$

This implies $x_w \in \text{int } B(x_0,r)$ because of (f) and

$$||x_w - x_0|| \leqq \frac{1}{\alpha}(L_r + \gamma)\delta(w,w_0)^{1/\beta}$$

In the finite dimensional case Theorem 6 can be improved as follows.

COROLLARY 2 In addition to the assumptions of Theorem 6, let X be finite dimensional and let the mappings f and g be continuous. Then there are reals $r > 0$, $\tilde{\alpha} > 0$, and a neighborhood V of w_0 such that the extremal value function μ_r is Lipschitz at w_0, and for each $w \in V$ the problem (P_w) has a local minimizer with

$$||x_w - x_0|| \leqq \tilde{\alpha}\delta(w,w_0)^{1/\beta}$$

PROOF. Let r and V be determined as in Theorem 6. Then $\Sigma_r(w) \neq \emptyset$ for all $w \in V$. In addition, $\Sigma_r(w)$ is closed and bounded. Since f is continuous, the problem

$$\text{Minimize } f(x,w) \quad \text{subject to } x \in \Sigma_r(w) \qquad (\tilde{P}_w)$$

has an optimal solution x_w. By Theorem 6 it follows that $x_w \in \text{int } B(x_0 r)$. Hence x_w is a local minimizer of (P_w). The remaining assertions were proved in Theorem 6.

A more detailed discussion of finite dimensional problems can be found in Robinson [12]. Under similar hypotheses as in Theorem 6, Levitin [5] obtained a stronger result. However, an example given by Robinson [12] shows that this result cannot be correct.

REMARK 4. If the second order sufficient optimality condition of Maurer [8] is satisfied, then (B) holds with $\beta = 2$. In this case we obtain bounds of the form

$$||x_w - x_0|| \leqq \alpha\delta(w,w_0)^{1/2}$$

This bound can be improved if we impose stronger assumptions on f and g; for example,

$$||g(x,w) - g(x,w_0)|| \leqq L_g\delta(w,w_0)^2$$

instead of (3.4). However, it should be remarked that this condition is not satisfied for standard perturbations ($W = Y$, $g(x,w) = g(x) + w$).

V. APPENDIX: PROOF OF THEOREM 1

In this appendix we give a proof of Theorem 1 which is based on results presented in Alt [1].

THEOREM 7 (Alt [1], Theorem II.2.8) Let the assumptions of Theorem 1 be fulfilled. Then for any $\varepsilon > 0$ there exist neighborhoods $\tilde{U}_\varepsilon$ of x_0, $\tilde{V}_\varepsilon$ of w_0, and D_ε of 0 in Y such that for each $w \in \tilde{V}_\varepsilon$ the set $\Sigma(w)$ is nonempty and for each such w we have

$$d[x,\Sigma(w)] \leqq (L + \varepsilon)||g(x,w) - g(x_0,w_0) - g_x(x_0,w_0)(x - x_0) - y|| \tag{5.1}$$

for all $(x,y) \in M_\varepsilon$, where $L > 0$ is a constant, which can be chosen independently of ε and

$$M_\varepsilon = \{(x,y) \in (C \cap \tilde{U}_\varepsilon) \times D_\varepsilon \mid g(x_0,w_0) + g_x(x_0,w_0)(x - x_0) + y \in K\}$$

COROLLARY 3 Let the assumptions of Theorem 1 be fulfilled. Then for any $\varepsilon > 0$ there exist neighborhoods U_ε of x_0 and V_ε of w_0 such that for each $w \in V_\varepsilon$ the set $\Sigma(w)$ is nonempty and for each such w (5.1) holds for all $(x,y) \in \tilde{M}_\varepsilon$, where

$$\tilde{M}_\varepsilon = \{(x,y) \in (C \cap U_\varepsilon) \times Y \mid g(x_0,w_0) + g_x(x_0,w_0)(x - x_0) + y \in K\}$$

PROOF. For any $\varepsilon > 0$ let the sets $\tilde{U}_\varepsilon$, $\tilde{V}_\varepsilon$, and D_ε be determined by Theorem 7. We choose neighborhoods U_ε of x_0, V_ε of w_0, and a real number $\delta > 0$ with the following properties:

$$U_\varepsilon \subset \tilde{U}_\varepsilon,\ V_\varepsilon \subset \tilde{V}_\varepsilon,\ B(0,\delta) \subset D_\varepsilon,$$

$$||g(x,w) - g(x_0,w_0)|| < \frac{\delta}{4} \qquad \text{for all } (x,w) \in U_\varepsilon \times V_\varepsilon$$

$$||g_x(x_0,w_0)(x - x_0)|| < \frac{\delta}{4} \qquad \text{for all } x \in U_\varepsilon \tag{5.2}$$

Then it follows from Theorem 7 that $\Sigma(w) \neq \emptyset$ for each $w \in V_\varepsilon$. Let $(x,w) \in U_\varepsilon \times V_\varepsilon$ be arbitrary and set $y_0 = -g_x(x_0,w_0)(x - x_0)$. By (5.2) we have $y \in D_\varepsilon$, $g(x_0,w_0) + g_x(x_0,w_0)(x - x_0) + y_0 = g(x_0,w_0) \in K$, and therefore

$$d[x,\Sigma(w)] \leqq (L + \varepsilon)||g(x,w) - g(x_0,w_0)|| \leqq (L + \varepsilon)\frac{\delta}{4} \tag{5.3}$$

by Theorem 7.

For arbitrary $y \in Y$ with $(x,y) \in \tilde{M}_\varepsilon$ we set

$$\bar{y} = g(x,w) - g(x_0,w_0) - g_x(x_0,w_0)(x - x_0) - y$$

If $||\bar{y}|| \geqq (\delta/2)$ then it follows from (5.3) that $d[x,\Sigma(w)] \leqq (L + \varepsilon)\frac{\delta}{4} < (L + \varepsilon)||\bar{y}||$, which implies (5.1). If $||\bar{y}|| < (\delta/2)$ then $||y|| \leqq ||g(x,w) - g(x_0,w_0)|| + ||g_x(x_0,w_0)(x - x_0)|| + ||\bar{y}|| < \delta$; hence $y \in D_\varepsilon$ and (5.1) holds by Theorem 7.

The proof of Theorem 1 can now easily be derived from Corollary 3.

PROOF OF THEOREM 1. Let the sets U_ε and V_ε be determined by Corollary 3. It remains to prove (2.4). For any $y \in G(x,w)$ there is some $k \in K$ with $y = g(x,w) - k$. We set

$$\bar{y} = g(x,w) - g(x_0,w_0) - g_x(x_0,w_0)(x - x_0) - y$$

Then we have $g(x_0,w_0) + g_x(x_0,w_0)(x - x_0) + \bar{y} = g(x,w) - y = k \in K$; hence Corollary 3 implies

$$\begin{aligned} d[x,\Sigma(w)] &\leqq (L + \varepsilon)||g(x,w) - g(x_0,w_0) - g_x(x_0,w_0)(x - x_0) - \bar{y}|| \\ &= (L + \varepsilon)||y|| \end{aligned}$$

Since $y \in G(x,w)$ was arbitrary, we conclude that

$$d[x,\Sigma(w)] \leqq (L + \varepsilon) \inf\{||y|| \mid y \in G(x,w)\} = (L + \varepsilon)d[0,G(x,w)]$$

This completes the proof.

REFERENCES

1. W. Alt. Stabilität mengenwertiger Abbildungen mit Anwendungen auf nichtlineare Optimierungsprobleme, *Bayreuther Mathematische Schriften*, Heft 3 (1979).
2. A. V. Fiacco and W. P. Hutzler. Basic results in the development of sensitivity and stability analysis in constrained mathematical programming, Report No. T-407, Institute for Management Science and Engineering, George Washington University (1979).
3. J. Gauvin and F. Dubeau. Differential properties of the marginal function in mathematical programming, to appear.
4. F. Lempio and H. Maurer. Differential stability in infinite-dimensional nonlinear programming, *Appl. Math. Optimization*, Vol. 6, pp. 139-152 (1980).
5. E. S. Levitin. On the local perturbation theory of a problem of mathematical programming in a Banach space, *Dokl. Akad. Nauk. SSSR*, Vol. 224, pp. 1354-1358 (1975).
6. E. S. Levitin. Differentiability with respect to a parameter of the optimal value in parametric problems of mathematical programming, *Kibernetika*, Vol. 1, pp. 44-59 (1976).
7. J. Maguregui. Regular multivalued functions and algorithmic applications. Ph.D. dissertation, University of Wisconsin-Madison, 1976.
8. H. Maurer and J. Zowe. First and second-order necessary and sufficient optimality conditions for infinite-dimensional programming problems, *Math. Programming*, Vol. 16, pp. 98-110 (1979).
9. S. M. Robinson. Stability theory for systems of inequalities, Part II:

Differentiable nonlinear systems, *SIAM J. Numer. Anal.*, Vol. 13, pp. 497-513 (1976).

10. S. M. Robinson. Regularity and stability for convex multi-valued functions, *Mathematics of Operations Research*, Vol. 1, pp. 597-607 (1976).

11. S. M. Robinson. Generalized equations and their solutions, Part I: Basic theory, *Mathematical Programming Study*, Vol. 10, pp. 128-141 (1979).

12. S. M. Robinson. Generalized equations and their solutions, Part II: Applications to nonlinear programming, Technical Summary Report No. 2048, Mathematics Research Center, University of Wisconsin-Madison (1980).

13. R. W. H. Sargent. On the parametric variation of constraint sets and solutions of minimization problems, to appear.

14. J. Zowe and S. Kurcyusz. Regularity and stability for the mathematical programming problem in Banach spaces, *Appl. Math. Optimization*, Vol. 5, pp. 49-62 (1979).

CHAPTER 3 ON THE CONTINUITY OF THE OPTIMUM SET IN PARAMETRIC SEMIINFINITE PROGRAMMING

BRUNO BROSOWSKI / Johann Wolfgang Goethe Universität, Frankfurt

I. INTRODUCTION AND NOTATION

Let T be a compact Hausdorff space, U a nonempty subset of $\mathbb{R}^N$, $A : T \times U \to \mathbb{R}$ and $p : U \to \mathbb{R}$ continuous mappings. For every continuous function b in C(T) we consider the following minimization problem:

$$\begin{aligned} &\text{Minimize } p : U \to \mathbb{R} \\ &\text{subject to } A(t,x) \leqq b(t) \qquad \text{for all } t \in T \end{aligned} \tag{*}$$

In this way a family of minimization problems is defined with family parameter b and parameter space C(T). For each element b in C(T) we define the set of feasible points

$$Z_b := \{x \in U \mid A(t,x) \leqq b(t) \text{ for all } t \in T\}$$

the minimum value

$$E_b := \inf\{p(v) \in \mathbb{R} \mid v \in Z_b\}$$

and the set of minimal points

$$P_b := \{v \in Z_b \mid p(v) = E_b\}$$

Further, we introduce the solution set

$$\mathsf{L}_{A,p} := \{b \in C(T) \mid P_b \neq \emptyset\}$$

Obviously, the sets Z_b and P_b and the real number E_b depend on the parameter b in $\mathsf{L}_{A,p}$.

In earlier papers [2,3,4,5,6] the author resp. K. Schnatz and the author derived sufficient and necessary conditions for the lower semicontinuity of the mapping

$$P : \mathsf{L}_{A,p} \to 2^U$$

However, in these papers it was assumed that the minimization problem is linear and that the mapping P is in addition upper semicontinuous and compact valued. Here we show that the conditions for lower semicontinuity are also true under weaker assumptions. In particular, we show in Section II that the sufficient conditions for lower semicontinuity of P [2,4] are still sufficient, if we only assume that P is closed and has locally a bounded selection. Further, we have assumed that the minimization problem is linear. But certain results of approximation theory indicate that this condition is also valid for nonlinear minimization problems with $A(\cdot,P_b)$ convex (see Reindlmeier [10]). In Section III we show that the sufficient condition of Theorem 2 is also necessary for lower semicontinuity. In this case, we do not have to assume that the minimization problem is linear; it suffices to assume that the sets $A(\cdot,P_b)$ are convex and finite dimensional. In Section IV we prove a necessary and sufficient condition for lower semicontinuity when T is in addition an m-dimensional compact C^K-manifold, where $1 \leqq K \leqq \infty$. Further, we assume that the parameter space is a subset of $C^s(T)$, denoting the space of all s times continuously differentiable functions on $T, 1 \leqq s \leqq K$. We also assume that the element $A(\cdot,v)$ is contained in $C^s(T)$ for all $v \in U$. In Section V we consider some applications. First we show that for a certain method the approximate solution of the Dirichlet problem does not depend continuously on the boundary values. Further, we give a sufficient condition for the lower semicontinuity of the metric projection onto linear subspaces in a space of vector valued functions.

We state necessary definitions. For each b in $\mathsf{L}_{A,p}$ and v in P_b let

$$M_{b,v} := \{t \in T \mid A(t,v) = b(t)\}$$

$$\mathsf{E}_b := \bigcap_{v \in P_b} M_{b,v}$$

and

$$\mathsf{N}_b := \bigcap_{v \in P_b} \{t \in T \mid A(t,v) = A(t,v_0)\}$$

where $v_0 \in P_b$ is arbitrary. It is clear that N_b does not depend on the choice of v_0 in P_b.

II. SUFFICIENT CONDITIONS FOR THE LOWER SEMICONTINUITY OF P

In this section we consider a more special minimization problem:

$$\begin{aligned} &\text{Minimize the linear functional} \quad p : \mathbb{R}^N \to \mathbb{R} \\ &\text{subject to } A(t) \cdot x \leqq b(t) \qquad \text{for all } t \in T \\ &\text{where } A : T \to \mathbb{R}^N \text{ is a continuous mapping} \end{aligned} \tag{LM}$$

We define for this special minimization problem sufficient conditions for the lower semicontinuity of the mapping

$$P : \mathsf{L}_{A,p} \to 2^{\mathbb{R}^N}$$

DEFINITION 1 The mapping

$$Z : \mathsf{L}_{A,p} \to 2^{\mathbb{R}^N}$$

is called a P-mapping in $b_0 \in \mathsf{L}_{A,p}$ iff for every z,x in Z_{b_0} there exist neighborhoods $U_x \subset \mathbb{R}^N$ and $U_{b_0} \subset \mathsf{L}_{A,p}$ and an $\alpha > 0$ such that $v + \alpha(z - x) \in Z_b$ for all b in U_{b_0} and for all v in $U_x \cap Z_b$.

THEOREM 1 Assume that for some $b_0 \in \mathsf{L}_{A,p}$ there exist a neighborhood $U_{b_0} \subset \mathsf{L}_{A,p}$ and a mapping $\pi : U_{b_0} \to \mathbb{R}^N$ such that

1. $\pi(b) \in P_b$ for all b in U_{b_0}
2. there exists a real number $K > 0$ such that $||\pi(b)|| \leqq K$ for all b in U_{b_0},

i.e., π is a bounded selection of P on U_{b_0}. If $Z : \mathsf{L}_{A,p} \to 2^{\mathbb{R}^N}$ is a P-mapping in $b_0 \in \mathsf{L}_{A,p}$ and if P is a closed mapping, then the mapping

$$P : \mathsf{L}_{A,p} \to 2^{\mathbb{R}^N}$$

is lower semicontinuous in b_0.

PROOF. If P is not lower semicontinuous in b_0, there is a sequence (b_n) in $L_{A,p}$, an element $\bar{v}_0$ in P_{b_0}, and an open set W_0 containing $\bar{v}_0$ such that $b_n \to b_0$ and

$$P_{b_n} \cap W_0 = \emptyset \qquad \text{for all n in } \mathbb{N}$$

By assumption, we can choose for each $n \in \mathbb{N}$ an element v_n in P_{b_n} such that (v_n) is bounded. Then there is a subsequence of (v_n) [which we again denote by (v_n)] and an element $v_0 \in P_{b_0}$ such that

$$v_n \to v_0 \neq \bar{v}_0$$

For each $n \in \mathbb{N}$ and for each $\lambda \geqq 0$, define the element

$$w_n(\lambda) := v_n + \lambda(\bar{v}_0 - v_0)$$

Assume that there exists a sequence (λ_k) of real numbers such that

$$\lambda_k \to \lambda \geqq 1$$

and a subsequence $(b_{n(k)})$ such that

$$w_{n(k)}(\lambda_k) \in P_{b_{n(k)}}$$

The sequence $(w_{n(k)}(\lambda_k))$ converges to an element

$$w := v_0 + \lambda(\bar{v}_0 - v_0)$$

Since $\lambda \geqq 1$, there exists a real number $\mu \in [0,1]$ such that

$$\bar{v}_0 = \mu v_0 + (1 - \mu)w$$

The sequence

$$\mu v_{n(k)} + (1 - \mu)w_{n(k)}(\lambda_k) \in P_{b_{n(k)}}$$

converges to the element $\bar{v}_0$, which contradicts

$$P_{b_{n(k)}} \cap W_0 = \emptyset$$

for $k \in \mathbb{N}$. Consequently, for each $n \geqq n_1$ $(n_1 \in \mathbb{N})$ there is a $1 > \lambda_n \geqq 0$ such that

$$w_n := v_n + \lambda_n(\bar{v}_0 - v_0) \in P_{b_n}$$

and

$$v_n + \lambda(\bar{v}_0 - v_0) \notin P_{b_n} \tag{*}$$

for $\lambda > \lambda_n$. Since $v_n \in P_{b_n}$ and $p(v_n) = p(v_n + \lambda(\bar{v}_0 - v_0))$ for every λ we can conclude from (*) also that

$$v_n + \lambda(\bar{v}_0 - v_0) \notin Z_{b_n}$$

for $\lambda > \lambda_n$. As before, one can prove that

$$\lambda_0 := \sup\{\lambda_n \in [0,1]\} < 1$$

which implies that the element

$$w := v_0 + \lambda_0(\bar{v}_0 - v_0)$$

is contained in P_{b_0}. Now we have

$$\begin{aligned} w_n + \frac{1}{n}(\bar{v}_0 - w) &= v_n + \lambda_n(\bar{v}_0 - v_0) + \frac{1}{n}\bar{v}_0 - \frac{1}{n}v_0 - \frac{\lambda_0}{n}(\bar{v}_0 - v_0) \\ &= v_n + \left[\lambda_n + \frac{1-\lambda_0}{n}\right](\bar{v}_0 - v_0) \end{aligned}$$

Since λ_n is maximal and since $\lambda_n + (1 - \lambda_0)/n > \lambda_n$, the element $w_n + (1/n)(\bar{v}_0 - w)$ is not contained in Z_{b_n}. But this is impossible, since Z is a P-mapping in b_0. The theorem is thus proved.

REMARK 1. If $|T| < \infty$, or if the mapping $P : \mathsf{L}_{A,p} \to 2^{\mathbb{R}^N}$ is upper semicontinuous and compact valued, then P is a closed mapping and has a locally bounded selection. Using this remark we can prove

COROLLARY 1 Let the minimization problem (LM) be given and assume that $|T| < \infty$. Then the mapping

$$P : \mathsf{L}_{A,p} \to 2^{\mathbb{R}^N}$$

is lower semicontinuous.

PROOF. It suffices to show that Z is a P-mapping. Let 1,2,...,m be the elements of T and let the elements $b_0 \in \mathsf{L}_{A,p}$ and $x,z \in Z_{b_0}$ be given. Further, denote by I the set of indices $\mu \in \{1,2,\ldots,m\}$ such that

$$\ell_\mu := A(\mu) \cdot x < b_0(\mu)$$

Now we define

$$U_{b_0} := \left\{ b \in L_{A,p} \mid \forall_{\mu \in I} \frac{\ell_\mu + b_0(\mu)}{2} < b(\mu) \right\}$$

$$\varepsilon := \frac{1}{4} \min_{\mu \in I} [b_0(\mu) - \ell_\mu]$$

$$U_x := \left\{ y \in \mathbb{R}^N \mid ||y - x||_1 < \frac{\varepsilon}{\max|A(\mu)|} \right\}$$

and

$$\alpha := \min_{\mu \in I} \frac{\frac{1}{2}(b_0(\mu) - \ell_\mu) - \varepsilon}{b_0(\mu) - \ell_\mu}$$

For $\mu \in I$, $b \in U_{b_0}$, and $y \in U_x \cap Z_b$ we have

$$\begin{aligned} A(\mu) \cdot (y + \alpha(z - x)) &= A(\mu)x + A(\mu)(y - x) + \alpha A(\mu)(z - x) \\ &\leqq \ell_\mu + \varepsilon + \alpha(b_0(\mu) - \ell_\mu) \\ &\leqq \ell_\mu + \varepsilon + \frac{1}{2}(b_0(\mu) - \ell_\mu) - \varepsilon \\ &\leqq \frac{1}{2}(\ell_\mu + b_0(\mu)) < b(\mu) \end{aligned}$$

For $\mu \notin I$, $b \in U_{b_0}$, and $y \in U_x \cap Z_b$ we have

$$A(\mu)(y + \alpha(z - x)) \leqq b(\mu) + \alpha(b_0(\mu) - b_0(\mu)) = b(\mu)$$

Consequently $y + \alpha(z - x)$ in Z_b. Using Theorem 1 we can conclude Corollary 1.

REMARK 2. Corollary 1 is well known. It was first established by Böhm [1] and Reindlmeier [10]. Their proofs use different methods.

COROLLARY 2 Let the minimization problem (LM) be given and assume that $|T| < \infty$. Then the mapping

$$Z : L_{A,p} \to 2^{\mathbb{R}^N}$$

is lower semicontinuous.

PROOF. Choose for p the zero functional and use Corollary 1.

Another sufficient condition for lower semicontinuity is given by

THEOREM 2 Assume that for some $b_0 \in L_{A,p}$ there exist a neighborhood

$U_{b_0} \subset \mathsf{L}_{A,p}$ and a mapping $\pi : U_{b_0} \to \mathbb{R}^N$ such that

1. $\pi(b) \in P_b$ for all b in U_{b_0}
2. there exists a real number $K > 0$ such that $||\pi(b)|| \leq K$ for all b in U_{b_0}.

If the mapping $P : \mathsf{L}_{A,p} \to 2^{\mathbb{R}^N}$ is closed and if there exists a nonempty open set $W \subset T$ such that

$$\mathsf{N}_b \supset W \supset \mathsf{E}_b$$

then the mapping P is lower semicontinuous at b.

PROOF. We assume that P is not lower semicontinuous in b. Then there exists a sequence (b_n) of elements b_n in $\mathsf{L}_{A,p}$ an element v_0 in P_b and a neighborhood U_0 of v_0 such that (b_n) converges to b and

$$P_{b_n} \cap U_0 = \emptyset$$

for each $n \in \mathbb{N}$. We can assume without loss of generality that v_0 is a relative interior point of P_b. By assumption, we can choose for each $n \in \mathbb{N}$ an element v_n in P_{b_n} such that (v_n) is bounded. Then there is a subsequence of (v_n) [which we again denote by (v_n)] and an element $v' \in P_{b_0}$ such that $v_n \to v' \neq \bar{v}_0$. Then it follows that

$$\mathsf{E}_b = \bigcap_{v \in P_b} M_{b,v} = M_{b,v_0}$$

Note that the sequence $\bar{v}_n := v_0 + v_n - v'$ converges to v_0. Thus for $n \geq n_1$ (for some $n_1 \in \mathbb{N}$) we have:

$$\bar{v}_n \notin P_{b_n}$$

but also $p(\bar{v}_n) = p(v_n) = E_{b_n}$. It follows that

$$\bar{v}_n \notin Z_{b_n}$$

or

$$\max_{t \in T} (A(t) \cdot \bar{v}_n - b_n(t)) > 0$$

for all $n \geq n_1$. If we abbreviate

$$H_n(t) := A(t) \cdot \bar{v}_n - b_n(t)$$

and

$$R_n := \left\{ t \in T \mid H_n(t) = \max_{\tau \in T} H_n(\tau) \right\}$$

then we get for each $t \in R_n$:

$$0 < H_n(t) = A(t) \cdot (v_0 - v') + A(t) \cdot v_n - b_n(t) \leqq A(t) \cdot (v_0 - v')$$

which yields

$$N_b \cap R_n = \emptyset \qquad \text{for all } n \geqq n_1$$

and in particular

$$W \cap R_n = \emptyset \qquad \text{for all } n \geqq n_1$$

On the other hand, since W is open and $E_b \subset W$, it follows that

$$\sup_{t \in T \setminus W} (A(t) \cdot v_0 - b(t)) < 0$$

Hence there exists a number $\varepsilon > 0$ such that for all $t \in T \setminus W$,

$$A(t) \cdot v_0 - b(t) \leqq -2\varepsilon$$

Note that the sequence (H_n) converges to the element $A \cdot v_0 - b$, so for sufficiently large n and for all $t \in T \setminus W$ we have

$$H_n(t) \leqq -\varepsilon < 0$$

which yields the inclusion:

$$T \setminus W \subset T \setminus R_n \qquad \text{or} \qquad W \supset R_n$$

but this contradicts $W \cap R_n = \emptyset$. Thus the theorem is proved.

REMARK 3. Since every subset of a finite subset is open and closed, Theorem 2 implies immediately Corollary 1.

III. A NECESSARY CONDITION FOR THE LOWER SEMICONTINUITY OF P

In this section we show that the sufficient condition of Theorem 2 is also necessary for subsets $B \subset L_{A,p}$ with the following additional property:

DEFINITION 2 A nonempty subset $B \subset L_{A,p}$ is said to have property (C) if $b_0 \in B$ and $P_{b_0} \supset P_b$ implies $b \in B$.

It is remarkable that we do not have to assume that the minimization problem (*) is linear; it suffices to assume that the sets $A(\cdot,P_b)$ are convex and finite dimensional.

We now prove the

THEOREM 3 Let the minimization problem (*) be given and let $\mathcal{B} \subset \mathcal{L}_{A,p}$ be a nonempty subset such that

1. $\mathcal{B}$ has property (C),
2. $A(\cdot,P_b)$ is convex and finite dimensional for each $b \in \mathcal{B}$.

If the mapping $P : \mathcal{B} \to 2^U$ is lower semicontinuous, then for every b in $\mathcal{B}$ the closed set $\mathcal{N}_b$ is also open.

PROOF. The lower semicontinuity of the mapping $P : \mathcal{B} \to 2^U$ implies that also the set valued mapping

$$Q : b \to A(\cdot,P_b)$$

is lower semicontinuous. Suppose now that the theorem is false; that is, there exists an element b_1 in $\mathcal{B}$ such that $\mathcal{N}_{b_1}$ is not open. The set Q_{b_1} consists of more than one point, since otherwise $\mathcal{N}_{b_1} = T$ would be open. Since Q_{b_1} is convex and finite dimensional, there are elements $v_0,v_1,\ldots,v_k$ in P_{b_1} such that

$$A(\cdot,v_1),\ A(\cdot,v_2),\ \ldots,\ A(\cdot,v_k)$$

are linearly independent, $A(\cdot,v_0) \neq A(\cdot,v_\kappa)$ for all $\kappa = 1,2,\ldots,k$, and k maximal. Since $\mathcal{N}_{b_1}$ is not open, there exists a point $t_0 \in \mathcal{N}_{b_1}$ with the following property:

every neighborhood U of t_0 contains some point t_U such that $A(t_U,v_\kappa) \neq A(t_U,v_0)$ (*)

for at least one $\kappa \in \{1,2,\ldots,k\}$.

Now we construct an element b in $\mathcal{B}$ as follows: In the case where t_0 is in M_{b_1,v_0}, we define $b := b_1$; if t_0 is not in M_{b_1,v_0} then we choose a neighborhood U' of t_0 such that

$$M_{b_1,v_0} \cap U' = \emptyset$$

Since $A(t_0,v_\kappa) = A(t_0,v_0)$ for all $\kappa = 1,2,\ldots,k$, and $b_1(t_0) - A(t_0,v_0) > 0$, there is an $\alpha > 0$ such that for all t in some neighborhood $U \subset U'$ of t_o we have:

$$b_1(t) - A(t,v_0) \geqq \alpha > 0$$

and

$$H^2(t) := \sum_{\kappa=1}^{k} (A(t,v_\kappa) - A(t,v_0))^2 \leqq \alpha^2$$

with $H(t) \geqq 0$. By Urysohn's lemma, there is a continuous function $\rho : T \to [0,1]$ with $\rho(t_0) = 1$ and $\rho(t) = 0$ off U. It we let

$$g(t) := \begin{cases} 1 - H(t)/(b_1(t) - A(t,v_0)) & \text{for } t \in U \\ 0 & \text{otherwise} \end{cases}$$

then for $s(t) := g(t) \cdot \rho(t)$, we have $s \in C[T]$, $0 \leqq s(t) \leqq 1$ on T, $s(t) = 0$ off U, and $s(t_0) = 1$. Now we define

$$b(t) := (1 - s(t)) \cdot b_1(t) + s(t) \cdot A(t,v_0)$$

The function b has the following properties:

1. $v_0 \in P_b$,
2. $P_b \subset P_{b_1}$, and consequently $b \in \mathcal{B}$,
3. $v_\kappa \in P_b$, $\kappa = 1,2,\ldots,k$.

Proof of the properties:

1. The element v_0 satisfies the inequality $A(t,v_0) \leqq b_1(t)$. From the follows $(1 - s(t))A(t,v_0) \leqq (1 - s(t))b_1(t)$, and by adding $s(t) \cdot A(t,v_0)$, we obtain $A(t,v_0) \leqq (1 - s(t))b_1(t) + s(t) \cdot A(t,v_0) = b(t)$; consequently, $v_0 \in Z_b$. Since $Z_b \subset Z_{b_1}$, it follows that

$$p(v_0) = \inf_{v \in Z_{b_1}} p(v) \leqq \inf_{v \in Z_b} p(v) \leqq p(v_0)$$

Hence v_0 is in P_b.

2. For each v in P_b we have $v \in Z_b \subset Z_{b_1}$, and thus $p(v) = p(v_0) = E_{b_1}$. Hence $v \in P_{b_1}$.

3. Since $v_0 \in P_b$ and $p(v_\kappa) = p(v_0)$, $\kappa = 1,2,\ldots,k$, it suffices to show that $v_\kappa \in Z_b$, i.e.,

$$\forall_{t \in T} \; A(t,v_\kappa) \leqq b(t)$$

Since $b_1(t) = b(t)$ for all t in $T \setminus U$, and $v_\kappa \in P_{b_1}$, $\kappa = 1,2,\ldots,k$, we need only consider the case $t \in U$. In this case for arbitrary $\kappa_0 \in \{1,2,\ldots,k\}$

we have

$$\begin{aligned} A(t,v_{\kappa_0}) - b(t) &= A(t,v_{\kappa_0}) - (1 - s(t))b_1(t) - s(t)A(t,v_0) \\ &= A(t,v_{\kappa_0}) - b_1(t) + s(t)(b_1(t) - A(t,v_0)) \\ &\leqq A(t,v_{\kappa_0}) - b_1(t) + b_1(t) - A(t,v_0) - H(t) \\ &= A(t,v_{\kappa_0}) - A(t,v_0) - H(t) \\ &\leqq A(t,v_{\kappa_0}) - A(t,v_0) - |A(t,\kappa_0) - A(t,v_0)| \leqq 0 \end{aligned}$$

Hence $v_1, v_2, \ldots, v_k$ are in P_b and consequently also $A(\cdot,v_1), A(\cdot,v_2), \ldots,$ $A(\cdot,v_k)$ are in Q_b.

Since $A(\cdot,v_0)$ is also in Q_b, there is an element $\bar{v} \varepsilon P_b$ such that

$$A(\cdot,\bar{v}) = \frac{1}{k+1} \sum_{\kappa=0}^{k} A(\cdot,v_\kappa)$$

is a relative interior point of the convex set Q_b. Then we have the relation

$$M_{b,v} = E_b \subset N_b = N_{b_1}$$

Since $E_b \subset N_b$ is trivial, we prove only $M_{b,\bar{v}} = E_b$. Let v be a point in P_b such that $A(\cdot,v) \neq A(\cdot,\bar{v})$. Then one can find a point $A(\cdot,v_1)$ in Q_b and a real number ρ with $0 < \rho < 1$ such that

$$A(\cdot,\bar{v}) = \rho A(\cdot,v_1) + (1 - \rho)A(\cdot,v)$$

Now let $M_{b,\bar{v}}$. Then we have

$$\begin{aligned} 0 &= A(t,\bar{v}) - b(t) \\ &= \rho A(t,v_1) + (1 - \rho)A(t,v) - b(t) \\ &\leqq (1 - \rho)(A(t,v) - b(t)) \end{aligned}$$

and consequently

$$0 \leqq A(t,v) - b(t)$$

Since we also have

$$A(t,v) - b(t) \leqq 0$$

we can conclude that t in $M_{b,v}$; hence we have the inclusion $M_{b,v} \supset M_{b,\bar{v}}$

for each $v \in P_b$. This implies

$$E_b = \bigcap_{v \in P_b} M_{b,v} = M_{b,\bar{v}}$$

Since t_0 is not an interior point of N_b, there is a net $(t_\lambda : \lambda \in \Lambda)$ of points t_λ in T such that (t_λ) converges to t_0 and, for each λ, $A(t_\lambda, v_\kappa) \neq A(t_\lambda, \bar{v})$, at least for one $\kappa \in \{1,2,\dots,k\}$. Then there exists an index κ_0 and a subnet $(t_\lambda : \lambda \in \Lambda_1)$ such that $A(t_\lambda, v_{\kappa_0}) \neq A(t_\lambda, \bar{v})$ for all t_λ with $\lambda \in \Lambda_1$. We may assume $\kappa_0 = 1$. By passing to a subnet $(t_\lambda : \lambda \in \Lambda_2)$ of $(t_\lambda : \lambda \in \Lambda_1)$, we can ensure that there exist signs $\varepsilon_\kappa \in \{-1,+1\}$ such that the inequalities

$$\varepsilon_1(A(t_\lambda, v_1) - A(t_\lambda, \bar{v})) < 0$$

and

$$\varepsilon_\kappa(A(t_\lambda, v_\kappa) - A(t_\lambda, \bar{v})) \leqq 0$$

for $\kappa = 2,3,\dots,k$ hold.

For each $\delta > 0$, the set

$$M_\delta := \left\{t \in T \mid |b(t_0) - b(t)| < \frac{\delta}{2}\right\}$$
$$\cap \left\{t \in T \mid |A(t_0, \bar{v}) - A(t, \bar{v})| < \frac{\delta}{2}\right\}$$

is a neighborhood of t_0. Hence there exists a $\lambda \in \Lambda_2$ such that $t_\lambda \in M_\delta$. Since $M_{b,\bar{v}} \subset N_b$ and $t_\lambda \notin N_b$, it follows that $t_\lambda \notin M_{b,\bar{v}}$. Since $M_{b,\bar{v}}$ is closed, there exists a compact neighborhood W of t_λ such that $M_{b,\bar{v}} \cap W = \emptyset$. Without loss of generality, we may assume $W \subset M_\delta$. Now let ρ be a continuous function such that $0 \leqq \rho(t) \leqq 1$ for $t \in T$, $\rho(t_\lambda) = 1$, $\rho(t) = 0$, for $t \in T \setminus W$, and define:

$$b_\delta(t) := (1 - \rho(t))b(t) + \rho(t)A(t, \bar{v})$$

This function b_δ has the following properties:

1. $\bar{v} \in P_{b_\delta}$,
2. $P_{b_\delta} \subset P_b$, and consequently $b_\delta \in B$,
3. $||b_\delta - b||_\infty < \delta$.

Proof of the properties:

The proofs of 1 and 2 are analogous to the proofs of the properties 1 and 2 for the function b, and therefore left to the reader. Property 3 is proved as follows:

For each $t \in W$ we have

$$\begin{aligned}|b_\delta(t) - b(t)| &= \rho(t)|b(t) - A(t,\bar{v})| \\ &= \rho(t)|b(t) - b(t_0) + A(t_0,\bar{v}) - A(t,\bar{v})| \\ &\leqq |b(t) - b(t_0)| + |A(t_0,\bar{v}) - A(t,\bar{v})| \\ &< \frac{\delta}{2} + \frac{\delta}{2} = \delta\end{aligned}$$

Since $|b_\delta(t) - b(t)| = 0$ for each $t \in T\backslash W$, property 3 follows.

For each element u in the set

$$G := \left\{A(\cdot,\bar{v}) + \sum_{\kappa=1}^{k} \varepsilon_\kappa \alpha_\kappa (A(\cdot,v_\kappa) - A(\cdot,\bar{v})) \right.$$
$$\left. \in C(T) \mid \alpha_\kappa < 0 \text{ for } \kappa = 1,2,\ldots,k\right\}$$

it follows that

$$u(t_\lambda) - b_\delta(t_\lambda) = A(t_\lambda,\bar{v}) + \sum_{\kappa=1}^{k} \varepsilon_\kappa\alpha_\kappa(A(t_\lambda,v_\kappa) - A(t_\lambda,\bar{v})) - A(t_\lambda,\bar{v})$$
$$= \sum_{\kappa=1}^{k} \varepsilon_\kappa\alpha_\kappa(A(t_\lambda,v_\kappa) - A(t_\lambda,\bar{v})) > 0$$

and consequently $u \notin A(\cdot,Z_{b_\delta})$, or $u \notin A(\cdot,P_{b_\delta}) = Q_{b_\delta}$. If we now let Y be the complementary space of span $\{A(\cdot,v_1), A(\cdot,v_2), \ldots, A(\cdot,v_k)\}$ in C(T), then

$$G' := G + Y$$

is an open set in C(T). Since

$$Q_{b_\delta} \subset A(\cdot,v_0) + \text{span}\,\{A(\cdot,v_1), A(\cdot,v_2), \ldots, A(\cdot,v_k)\}$$

we can conclude that

$$Q_{b_\delta} \cap G' = \emptyset$$

for all $\delta > 0$. The set G contains $A(\cdot,\bar{v})$ in its boundary, but $A(\cdot,\bar{v})$ is also a relative interior point of Q_b, so $Q_b \cap G' \neq \emptyset$. On the other hand, for each $\delta > 0$ there exists a b_δ in $\mathcal{B}$ with $||b_\delta - b|| < \delta$ and $Q_{b_\delta} \cap G' = \emptyset$.

This contradicts the lower semicontinuity of Q and hence also of P. Thus the theorem is proved.

IV. DIFFERENTIABLE PARAMETER FUNCTIONS

In this section we prove a necessary and sufficient condition for the case when T is in addition an m-dimensional compact C^K-manifold, where $1 \leqq K \leqq \infty$. These results extend earlier results of the author [5] and Schnatz and the author [6]. We first state some necessary definitions. Let $C^s(T)$ denote the space of all s times continuously differentiable functions on T. We assume that $A(\cdot,v)$ is contained in $C^s(T)$ for all v in U. For $b \in C^s(T)$ and $t \in T$, we denote by grad b(t) the vector of the first partial derivatives of b in t. For $b \in \mathbb{L}_{A,P}$ we define:

$$\mathbb{D}_b := \bigcap_{v \in P_b} \{t \in T \mid A(t,v) = A(t,v_0) \text{ and } \operatorname{grad} A(t,v) = \operatorname{grad} A(t,v_0)\}$$

where v_0 is an arbitrary element in P_b. It is obvious that $\mathbb{D}_b$ is independent of the choice of v_0 in P_b and also independent of the local coordinate neighborhood.

Let $\phi : T \to \mathbb{R}^m$ be a coordinate diffeomorphism and $t_0 \in \mathbb{D}_b$, i.e., for all $v \in P_b$ we have

$$A(t_0,v) = A(t_0,v_0) = b(t_0)$$

We denote by

$$e_\mu := (\delta_{\mu 1},\ldots,\delta_{\mu m}) \in \mathbb{R}^m \qquad \mu = 1,2,\ldots,m$$

the usual basis in $\mathbb{R}^m$. Then for sufficiently small $\Delta t \neq 0$ in $\mathbb{R}$, we have

$$A(\phi^{-1}(\phi(t_0) + \Delta t \cdot e_\mu),v) - A(t_0,v) \leqq b(\phi^{-1}(\phi(t_0 + \Delta t \cdot e_\mu)) - b(t_0)$$

$\mu = 1,2,\ldots,m$. Dividing both sides by Δt and letting $\Delta t \to 0$, we get

$$\frac{\partial}{\partial t_\mu} A(t_0,v) \leqq \frac{\partial}{\partial t_\mu} b(t_0) \qquad \text{as } \Delta t \to 0_+$$

and

$$\frac{\partial}{\partial t_\mu} A(t_0,v) \geqq \frac{\partial}{\partial t_\mu} b(t_0) \qquad \text{as } \Delta t \to 0_-$$

Thus since $A(\cdot,v)$ is assumed to be differentiable, for all v and v_0 in P_b we have

$$\operatorname{grad} A(t_0,v) = \operatorname{grad} A(t_0,v_0)$$

and hence the inclusion

$$\mathbb{N}_b \supset \mathbb{D}_b \supset \mathbb{E}_b$$

Now let the linear minimization problem (LM) be given. With the aid of Theorem 2, we can conclude the following sufficient condition for the lower semicontinuity of the mapping

$$P : \mathbb{L}_{A,p} \cap C^s(T) \to 2^{\mathbb{R}^N}$$

THEOREM 4 Assume that for each $b_0 \in \mathbb{L}_{A,p} \cap C^s(T)$ there exists a neighborhood $U_{b_0} \subset \mathbb{L}_{A,p} \cap C^s(T)$ and a mapping $\pi : U_{b_0} \to \mathbb{L}_{A,p}$ such that

1. $\pi(b) \in P_b$ for all b in U_{b_0},
2. there exists a real number $K > 0$ such that $||\pi(b)|| \leqq K$ for all b in U_{b_0}.

If the mapping $P : \mathbb{L}_{A,p} \cap C^s(T) \to 2^{\mathbb{R}^N}$ is closed and the closed set $\mathbb{D}_b$ is also open, then the mapping P is also lower semicontinuous at b.

Extending the methods developed in [4,5,6], we show that this criterion is also necessary. As in the corresponding theorem in Section III, we do not have to assume that the minimization problem (*) is linear; as in Section III it suffices to assume that the sets $A(\cdot,P_b)$ are convex and finite dimensional. We now prove

THEOREM 5 Let the minimization (*) be given and let $\mathbb{B} \subset \mathbb{L}_{A,p} \cap C^s(T)$ be a nonempty subset such that

1. $\mathbb{B}$ has property (C),
2. $A(\cdot,P_b)$ is convex and finite dimensional for each $b \in \mathbb{B}$.

If the mapping $P : \mathbb{B} \to 2^{\mathbb{R}^N}$ is lower semicontinuous, then for every $b \in \mathbb{B}$ the closed set $\mathbb{D}_b$ is also open.

PROOF. The lower semicontinuity of the mapping $P : \mathbb{B} \to 2^{\mathbb{R}^N}$ implies that also the set valued mapping

$$Q : b \to A(\cdot,P_b)$$

is lower semicontinuous. Suppose now that the theorem is false, that is, there exists an element b_1 in $\mathbb{B}$ such that $\mathbb{D}_{b_1}$ is not open. The set Q_{b_1} consists of more than one point, since otherwise $\mathbb{D}_{b_1} = T$ would be open. Since Q_{b_1} is convex and finite dimensional, there are elements $v_0,v_1,\ldots,v_k$ in P_{b_1} such that

$$A(\cdot,v_1),\ A(\cdot,v_2),\ \ldots,\ A(\cdot,v_k)$$

are linearly independent, $A(\cdot,v_0) \neq A(\cdot,v_\kappa)$ for $\kappa = 1,2,\ldots,k$, and k maximal. Since $\mathfrak{D}_{b_1}$ is open, there exists a point $t_0 \in \mathfrak{D}_{b_1}$ and a sequence (t_n) converging to t_0 such that for each $n \in \mathbb{N}$

1. $A(t_n,v) \neq A(t_n,v_0)$, or
2. $\operatorname{grad} A(t_n,v) \neq \operatorname{grad} A(t_n,v_0)$

for at least one $\kappa \in \{1,2,\ldots,k\}$. By choosing a subsequence of (t_n), if necessary, we can ensure that (t_n) is contained in a coordinate neighborhood of t_0.

Now we construct an element $b \in \mathfrak{B}$ as follows. In the case t_0 is in M_{b_1,v_0}, we define $b := b_1$. If t_0 is not in M_{b_1,v_0}, we choose a relatively compact coordinate neighborhood U of t_0 such that

$$M_{b_1,v_0} \cap \bar{U} = \emptyset.$$

Since $t_0 \in \mathfrak{D}_{b_1} \setminus M_{b_1,v_0}$ we have $A(t_0,v_\kappa) = A(t_0,v_0)$ and $\operatorname{grad} A(t_0,v_\kappa) = \operatorname{grad} A(t_0,v_0)$ for $\kappa = 1,2,\ldots,k$ and

$$b_1(t_0) - A(t_0,v_0) > 0$$

By reducing U, if necessary, we can ensure that for all $t \in U$,

$$b_1(t) - A(t,v_0) \geqq \alpha > 0$$

In the case $s = 1$ we can ensure in addition that

$$0 < H(t)^2 := \sum_{\kappa=1}^{k} (A(t,v_\kappa) - A(t,v_0))^2 \leqq \frac{1}{2} (\alpha \cdot r)^2$$

for suitable real numbers $0 < r \leqq 1$ and α with $0 < \alpha r < 1$. To treat the case $s > 1$, we need some further aids. If $\phi : V \to \mathbb{R}^m$ is a coordinate diffeomorphism, then we denote by

$$\frac{\partial^2}{\partial t_i \partial t_j} A(t_0,v_\kappa) \qquad i,j = 1,2,\ldots,m$$

the second partial derivatives of the function

$$A(\cdot,v\ )\circ\phi^{-1} : \mathbb{R}^m \to \mathbb{R}$$

at the point $x = x_0$, where $\phi(t) = (x_1,x_2,\ldots,x_m) = x$ and $\phi(t_0) = (x_1^0,x_2^0,\ldots,x_m^0) = x^0$. With this notation we define for $\kappa = 1,2,\ldots,k$ the (m,m)-matrix

$$A_\kappa(t_0) := \operatorname{diag}(-(2\alpha/d^2)) + H_\kappa$$

where the entries of H_κ in the ith row and jth column are

$$\frac{\partial^2}{\partial t_i \partial t_j} A(t_0, v_\kappa)$$

respectively. By Gerschgorin's lemma [11] all the eigenvalues of A_κ are contained in the sets $K_\mu := \{x \in \mathbb{R} \mid |x - x_\mu| < \rho_\mu\}$, $\mu = 1,2,\ldots,m$, where

$$x_\mu := -\frac{2\alpha}{d^2} + \frac{\partial^2}{\partial t_\mu^2} A(t_0, v_\kappa) \quad \text{and} \quad \rho_\mu := \sum_{j \neq \mu} \left| \frac{\partial^2}{\partial t_\mu \partial t_j} A(t_0, v_\kappa) \right|$$

If we choose $d > 0$ sufficiently small, we can ensure that all the eigenvalues of $A_\kappa(t_0)$ become negative; i.e., for $\kappa = 1,2,\ldots,k$ the matrices $A_\kappa(t_0)$ define negatively quadratic forms. Since the eigenvalues of a matrix depend continuously on its entries, there are neighborhoods U_κ, of t_0, $\kappa = 1, 2,\ldots,k$, such that for all $\tau \in U_\kappa$ the matrix $A_\kappa(\tau)$ has only negative eigenvalues. Reducing U if necessary, we can ensure that it is contained in $U_1 \cap U_2 \cap \ldots \cap U_k$.

Let $\phi : U \to \mathbb{R}^m$ be a coordinate diffeomorphism and $W := \{x \in \mathbb{R}^m \mid \|x - x_0\|_2 < d\}$, where $\|\cdot\|_2$ denotes the Euclidean norm in $\mathbb{R}^m$ and $x := \phi(t)$, $x_0 := \phi(t_0)$. For the case $s = 1$ we consider the mapping

$$h(x) := \begin{cases} 1 - \dfrac{\sqrt{H(\phi^{-1}(x)) + \frac{1}{2d^4} (\alpha r)^4 \|x - x_0\|_2^4}}{b(\phi^{-1}(x)) - A(\phi^{-1}(x), v_0)} & \text{if } x \in W \\ 0 & \text{if } x \notin W \end{cases}$$

and for the case $s > 1$,

$$h(x) := \begin{cases} 1 - \dfrac{\frac{\alpha}{d^2} \|x - x_0\|_2^2}{b_1(\phi^{-1}(x)) - A(\phi^{-1}(x), v_0)} & \text{for } x \in W \\ 0 & \text{for } x \notin W \end{cases}$$

In both cases we have $h \in C^s(W)$. Now we choose a function $g \in C^\infty(\mathbb{R}^m)$ such that

$$g(x) := \begin{cases} 1 & \text{for } \|x - x_0\|_2 \leqq (1/3)d \\ 0 & \text{for } \|x - x_0\| \geqq (2/3)d \end{cases}$$

and $0 \leqq g(x) \leqq 1$ for all $x \in \mathbb{R}^m$. Then the function $g \cdot h : \mathbb{R}^m \to \mathbb{R}$ is contained in $C^\infty(\mathbb{R}^m)$. By setting $h_1 := g \cdot h \cdot \phi$, we obtain a function h_1 in $C^s(T)$ with $0 \leqq h_1(t) \leqq 1$ for all $t \in T$, $h_1(t_0) = 1$, and $h_1(t) = 0$ for $t \in T \backslash U$. Now we define the function b by

$$b(t) := (1 - h_1(t))b_1(t) + h_1(t) \cdot A(t,v_0)$$

This function is contained in $C^s(T)$ and has the following properties:

1. $v_0 \in P_b$,
2. $P_b \subset P_{b_1}$, and consequently $b \in \mathcal{B}$,
3. $v_\kappa \in P_b$, $\kappa = 1,2,\ldots,k$.

The proofs of properties 1 and 2 are analogous to the ones given in Section III; also property 3 for the case $s = 1$ can be proved similarly. Thus it remains to show 3 for the case $s > 1$. Since $p(v_\kappa) = p(v_0)$, $\kappa = 1,2,\ldots,k$, it suffices to show that the elements v_κ are in Z_b, i.e., for all $t \in T$ we must show:

$$A(t,v_\kappa) \leqq b(t)$$

Since $b_1(t) = b(t)$ for all t in $T \backslash U$, we need only to consider the case $t \in U$. In this case for arbitrary $\kappa_0 \in \{1,2,\ldots,k\}$ we have:

$$\begin{aligned}
&A(t,v_{\kappa_0}) - b(t)\\
&= A(t,v_{\kappa_0}) - (1 - h_1(t))b_1(t) - h_1(t)A(t,v_0)\\
&= A(t,v_{\kappa_0}) - b_1(t) + h_1(t)(b_1(t) - A(t,v_0))\\
&\leqq A(t,v_{\kappa_0}) - b_1(t) + b_1(t) - A(t,v_0) - (\alpha/d^2)\ ||\phi(t) - \phi(t_0)||_2^2\\
&= A(t,v_{\kappa_0}) - A(t,v_0) - (\alpha/d^2)\ ||\phi(t) - \phi(t_0)||_2^2
\end{aligned}$$

Then Taylor expansion yields:

$$\begin{aligned}
A(t,v_{\kappa_0}) - b(t) \leqq{}& A(t_0,v_{\kappa_0}) - A(t_0,v_0) - (\alpha/d^2)\ ||\phi(t_0) - \phi(t_0)||_2^2\\
&+ (\phi(t) - \phi(t_0))\cdot\Big[\operatorname{grad}\Big(A(t,v_{\kappa_0}) - A(t_0,v_0)\\
&- \frac{\alpha}{d^2}||\phi(t)-\phi(t_0)||_2^2\Big)\Big]_{t=t_0}\\
&+ (\phi(t) - \phi(t_0)) \cdot A_\kappa(\tau) \cdot (\phi(t) - \phi(t_0)) \leqq 0
\end{aligned}$$

for some $\tau \in U$. This proves 3 for the case $s > 1$. Consequently, the elements $A(\cdot,v_1)$, $A(\cdot,v_2)$, ..., $A(\cdot,v_k)$ are in Q_b.

Since $A(\cdot,v_0)$ is also in Q_b, there is an element $\bar{v} \in P_b$ such that

$$A(\cdot,\bar{v}) = \frac{1}{k+1} \sum_{\kappa=0}^{k} A(\cdot,v_\kappa)$$

is a relative interior point of the convex set Q_b. As in Section III, one can prove the relation

$$M_{b,v} = E_b \subset N_b \subset N_{b_1}$$

Now we consider the sequence (t_n). If there is an infinite number of points t_n such that

$$A(t_n,v_\kappa) \neq A(t_n,v_0)$$

for some $\kappa \in \{1,2,\ldots,k\}$, we pass to a subsequence (t_n') such that

$$A(t_n',v_\kappa) \neq A(t_n',v_0)$$

for all $n \in \mathbb{N}$. We may assume $\kappa = 1$. If, on the other hand, for each $\kappa \in \{1,2,\ldots,k\}$ there is only a finite number of points t_n with $A(t_n,v_\kappa) \neq A(t_n,v_0)$, then we construct a new sequence (t_n') as follows. We may assume that

$$A(t_n,v_1) = A(t_n,v_0)$$

for each $n \in \mathbb{N}$. Then none of the points (t_n) is a local extremum of the function

$$A(\phi^{-1}(\cdot),v_1) = A(\phi^{-1}(\cdot),v_0)$$

in $\phi(U)$. Thus, for all $n \in \mathbb{N}$ there is, in the ball with center $\phi(t_n)$ and radius $1/n$, always an element c_n such that

$$A(\phi^{-1}(c_n),v_1) \neq A(\phi^{-1}(c_n),v_0)$$

Since D_b is a closed set, we can choose c_n such that $t_n' := \phi^{-1}(c_n)$ is not contained in D_b. In both cases the sequence (t_n') converges to t_0 and we have

$$A(t_n',v_1) \neq A(t_n',v_0)$$

semicontinuous, it follows that it is also continuous, and thus lower semicontinuous.

(2) Now assume that the mapping P is lower semicontinuous. Then, by Theorem 3, the closed set N_b is also open. Since p is not the zero functional and Z_b has a nonempty relative interior, it follows that N_b is nonempty. Since T is the disjoint sum of two connected spaces, it follows that

$$\mathsf{N}_b = \partial B_1, \ \mathsf{N}_b = \partial B_2, \text{ or } \mathsf{N}_b = \partial B_1 \cup \partial B_2$$

If $\mathsf{N}_b = \partial B_1$, then for some $(x_1^0, x_2^0, \ldots, x_N^0) \in P_b$ and for all $(x_1, x_2, \ldots, x_N) \in P_b$ and for all $t \in \partial B_1$, we have:

$$\sum_{\nu=1}^{N-1} (x_\nu - x_\nu^0) v_\nu(t) = 0$$

Since $v_1, v_2, \ldots, v_{N-1}$ are linearly independent, it follows that $x_\nu = x_\nu^0$, $\nu = 1,2,\ldots,N-1$. Since also $x_N = x_N^0 = E_b$, it follows that P_b consists of only one element. The other cases can be proved similarly.

It is well known that in the case $N \geqq 3$ the optimization problem (MD') is not uniquely solvable for each $b \in C(T)$. Thus, the approximate solution of the Dirichlet problem does not depend continuously on the boundary values.

The result of Theorem 4 remains true, even if we restrict the boundary values to differentiable functions:

THEOREM 7 The mapping $P : C^s(T) \to 2^{\mathbb{R}^N}$, $1 \leqq s \leqq \infty$, is lower semicontinuous if and only if the minimization problem (MD') is uniquely solvable.

PROOF. The proof is similar to the proof of Theorem 6. One has to use Theorem 5 instead of Theorem 3.

In contrast to the case of continuous function, there are linear subspaces

$$\text{span}(v_1, v_2, \ldots, v_{N-1}) \subset C^s(T)$$

with $N \geqq 3$ such that the optimization problem (MD') is uniquely solvable for all b in $C^s(T)$. A simple example is

$$v_1 = 1, \ v_2 = t_1, \ v_3 = t_2, \ v_4 = t_3$$

In this case the approximate solution depends, of course, continuously on the boundary values. It is of interest to determine all subspaces of $C^s(T)$ such that the optimization problem (MD') is uniquely solvable. A necessary and sufficient condition is given by

Each set of linearly independent functions
$u_1, u_2, \ldots, u_r \in \text{span}(v_1, v_2, \ldots, v_{N-1})$ has at most $N - 1 - r$ joint double zeroes in $T, r = 1,2,\ldots,N-1$ (UC)

Similar results are true, if one considers one-sided approximations to solutions of the Dirichlet problem.

B. Best Approximation of Vector-valued Functions

Let S be a compact Hausdorff space, and X be a finite dimensional strictly convex real normed linear space. We denote by C(S,X) the set of continuous mappings $b : S \to X$. If addition and multiplication with scalars are defined as usual, and the norm

$$||b|| := \max_{s \in S} ||b(s)||_X$$

is introduced, then C(S,X) is a normed linear space. Let $V \subset C(S,X)$ be a linear subspace with base $v_1, v_2, \ldots, v_{N-1}$. For any b in C(S,X), the set of best approximations of b from V is defined by

$$P_V(b) := \{v \in V \mid ||b - v|| = d(b,V)\}$$

where

$$d(b,V) := \inf\{||b - v|| \in \mathbb{R} \mid v \in V\}$$

The set-valued mapping $P_V : C(S,X) \to 2^V$ is called the metric projection onto V. This mapping is upper semicontinuous and compact valued. Further, we define the set

$$N_b := N_{b,V} := \bigcap_{v \in P_V(b)} \{t \in T \mid v(t) = v_0(t)\}$$

where v_0 is an arbitrary element in $P_V(b)$. Now we give a sufficient condition for the lower semicontinuity of P_V in C(S,X).

THEOREM 8 Let V be a finite dimensional linear subspace of the space $Z := C(S,X)$. Whenever for each b in Z the closed set N_b is also open, then the metric projection onto V is lower semicontinuous.

REMARK 4. This is a special case of a more general result of Wegmann and the author [7].

PROOF. We denote by E_{Z^*} the set of extreme points of the unit ball in the continuous dual Z* of Z. An element

$$v_0 = \sum_{\nu=1}^{N-1} x_\nu^0 v_\nu$$

is a best approximation of b from V if and only if $(x_1^0, x_2^0, \ldots, x_{N-1}^0, d(b,V))$ is a minimum point of the following minimization problem:

$$\text{minimize } p(x_1, x_2, \ldots, x_N) := x_N$$

$$\text{subject to } z^*\left(b - \sum_{\nu=1}^{N-1} x_\nu v_\nu\right) \leqq x_N \qquad \text{for all } z^* \text{ in } E_{Z^*} \qquad \text{(MA)}$$

Since the set E_{Z^*} is homeomorphic to the compact Hausdorff space $S \times E_{X^*}$, we have an optimization problem of type (LM) in Section II. The minimum set mapping P is connected with the metric projection P_V by the formula

$$P = P_V \times d(\cdot, V)$$

Since $d(\cdot,V)$ is always continuous, it follows that P is lower semicontinuous if and only if P_V is lower semicontinuous. It is easy to see that

$$M_{b,v} = \{z^* \in E_{Z^*} \mid z^*(b - v) = \|b - v\|\}$$

and

$$N_b = \bigcap_{v \in P_V(b)} \{z^* \in E_{Z^*} \mid z^*(v - v_0) = 0\}$$

where v_0 is an arbitrary element in $P_V(b)$. Since the closed set N_b is also open by hypothesis, the function g defined by

$$g(t) := \begin{cases} b(t) & \text{for } t \in N_b \\ 0 & \text{for } t \in T \setminus N_b \end{cases}$$

is in C(S,X). The set $W := \{z^* \in Z^* \mid z^*(g) > 0\}$ is a $\sigma(Z^*,Z)$-open subset of Z*. The functionals in E_{Z^*} are generalized evaluation functionals $\varepsilon_{x^*,s}$, that is, there exist elements $x^* \in E_{X^*}$ and $s \in S$ such that

$$\varepsilon_{x^*,s}(h) = x^*(h(s)) \qquad \text{for } h \in C(S,X)$$

By definition of W, for each functional $\varepsilon_{x^*,s} \in E_{Z^*} \cap W$, the equality

$$\varepsilon_{x^*,s}(v) = \varepsilon_{x^*}(v_0)$$

holds for all v in $P_V(b)$, whence we conclude

$$N_b \supset E_{Z*} \cap W$$

Now let $z^* \varepsilon E_b = \cap M_{b,v}$. For each v in $P_V(b)$, we have

$$\begin{aligned} d(b,V) &= z^*(b - v) \\ &= z^*(b(t) - v(t)) = ||b(t) - v(t)||_X \end{aligned}$$

Choose any v_0 in $P_V(b)$. Then for each v in $P_V(b)$ the element $(1/2)(v + v_0)$ is also in $P_V(b)$. Hence

$$||b(t) - \frac{1}{2}(v(t) + v_0(t))||_X = \frac{1}{2}||b(t) - v(t)||_X + \frac{1}{2}||b(t) - v_0(t)||_X$$

In view of the strict convexity of X, there is a positive real number μ such that

$$b(t) - v(t) = \mu(b(t) - v_0(t))$$

Using $||b(t) - v(t)|| = ||b(t) - v_0(t)||$, we obtain $\mu = 1$ and finally $v(t) - v_0(t) = 0$, whence $t \varepsilon N_b$, and consequently $z^* \varepsilon W \cap E_{Z*}$. Thus we have obtained the inclusion

$$N_b \supset E_{Z*} \cap W \supset E_b$$

Since the $E_{Z*} \cap W$ is open in the restriction of the weak * topology on E_{Z*}, the requirements of Theorem 2 are fulfilled, whence the lower semicontinuity of P_V follows.

It should be remarked that the condition of Theorem 8 is also necessary for the lower semicontinuity of P_V, even for certain nonlinear sets V. (Compare Reindlmeier [10].)

REFERENCES

1. V. Böhm. On the continuity of the optimal policy set for linear programs, *SIAM Journal of Applied Mathematics*, Vol. 28, pp. 303-306 (1975).

2. B. Brosowski. On parametric linear optimization, *Lecture Notes in Economics and Mathematical Systems*, Vol. 157, pp. 37-44 (1977).

3. B. Brosowski. Zur parametrischen linearen Optimierung: II, Eine hinreichende Bedingung fur die Unterhalbstetigkeit, *Operations Research Verfahren*, Vol. 31, pp. 137-141 (1979).

4. B. Brosowski. On parametric linear optimization: III, A necessary condition for lower semicontinuity, *Methods of Operations Research*, Vol. 36, pp. 21-30 (1980).

5. B. Brosowski. On parametric linear optimization: IV, Differentiable functions, *Lecture Notes in Economics and Mathematical Systems*, Vol. 179, pp. 31-39 (1980).

6. B. Brosowski and K. Schnatz. Parametric optimization differentiable parameter functions, *Methods of Operations Research*, Vol. 37, pp. 99-118 (1980).

7. B. Brosowski and R. Wegmann. On the lower semicontinuity of the set-valued metric projection, *Journal of Approximation Theory*, Vol. 8, pp. 84-100 (1973).

8. L. Collatz. Application of multivariate approximation to the solution of boundary value problems, in *Multivariate Approximation* (D. C. Handscomb, ed.), pp. 13-29. Academic Press, New York, 1978.

9. L. Collatz and W. Krabs. *Approximationstheorie*. B. G. Teubner-Verlag, Stuttgart, 1973.

10. J. Reindlmeier. Zur stetigen Parameterabhängigkeit der Menge der Minimalpunkte bei Minimaxaufgaben. Dissertation, Göttingen, 1975.

11. F. Stummel and K. Hainer. *Praktische Mathematik*. B. G. Teubner-Verlag, Stuttgart, 1971.

CHAPTER 4 OPTIMALITY CONDITIONS AND SHADOW PRICES*

HENRY WOLKOWICZ / The University of Alberta, Edmonton, Alberta, Canada

ABSTRACT

We consider an ordinary convex program that does not necessarily attain its infimum. It is well known that the Kuhn-Tucker multipliers, if they exist, provide shadow prices, or sensitivity coefficients, for the marginal improvement of the optimum with respect to perturbations in the right-hand sides of the constraints. Moreover, the Kuhn-Tucker multipliers exist if and only if this marginal improvement is bounded below for all perturbation directions. This is equivalent to "stability" of the program with respect to all non-negative right-hand side perturbations. We present a group of optimality conditions dependent on a regularizing set G_i. The "restricted" Kuhn-Tucker multipliers, thus obtained, provide shadow prices for unstable programs. Moreover, a class of perturbations is presented which, at the unit price given by the "restricted" Kuhn-Tucker multipliers, are never economical to buy, i.e., we find a stable subset of perturbations.

---

*Research partially supported by NSERC A3388.

I. INTRODUCTION

In this paper we consider the ordinary convex program

$$\begin{aligned} &\mu = \mathit{inf}\ p(x) \\ &\mathit{subject\ to}\ g^k(x) \leqq 0 \qquad k \in P = \{1,\dots,m\} \\ &x \in \Omega \end{aligned} \tag{P}$$

where p, $g^k : U \to R$ are convex functions on U, $U \subset R^n$ is some open set containing Ω, and Ω is a convex set.

If some constraint qualification, e.g., Slater's condition, holds for (P), then

$$\mu = \inf\{p(x) + \sum \lambda_k g^k(x) : x \in \Omega\} \tag{1}$$

for some Kuhn-Tucker multipliers $\lambda_k \geqq 0$. The λ_k provide shadow prices for (P), i.e., if

$$\mu(\varepsilon) = \inf\{p(x) : g^k(x) \leqq \varepsilon_k,\ k \in P,\ x \in \Omega\} \tag{2}$$

then, at the unit prices λ_k, no perturbation $\varepsilon = (\varepsilon_k)$ whatsoever is economical to buy. Moreover, this equilibrium situation characterizes the Kuhn-Tucker vector $\lambda = (\lambda_k)$, see, e.g., [11].

In the absence of any constraint qualification, no Kuhn-Tucker vector may exist. Characterizations of optimality without any constraint qualification, using feasible directions, have been given in, e.g., [1,2,3,4]. Stronger optimality conditions were given in [13,14,15,16]. These conditions all assumed that the infimum is attained and that $\Omega = R^n$.

In this paper we present several optimality conditions without constraint qualification of the type (1). In these conditions, Ω is replaced by $\Omega \cap G$, where G is some appropriate convex set. These results generalize the above mentioned characterizations, in the sense that the infimum need not be attained and Ω may be a proper, not necessarily polyhedral, subset of R^n. The proofs for these conditions are unified and depend on the characterization of optimality (with constraint qualification) given in Lemma 2 below. The unifying key to the proofs and method of choosing the sets G is that the equality set of constraints, denoted $P^=$, are affine on $\Omega \cap G$. (A generalization of the characterization of optimality in [1,2,4], to the case when the infimum may not be attained and $\Omega \neq R^n$, was first given in [5]. This generalization followed from a characterization of optimality for the abstract convex program.)

For these optimality conditions without constraint qualification, we find a set of perturbations $\varepsilon(G)$ which, at the unit prices λ_k, are never economical to buy, i.e., a stable set of perturbations. (Note that, in the case that the infimum is attained, the indices k such that the perturbation $t\varepsilon_k$ is always economical to buy for some $t > 0$, no matter what the cost, were found in [14]; see Section IV below.) We conclude with two examples.

II. PRELIMINARIES

Consider the ordinary convex program (P). Let $g(x) = (g^k(x))$ denote the vector of constraints in R^m. We assume that the *feasible set*

$$F = \{x \in \Omega : g(x) \leqq 0\} \neq \emptyset \tag{3}$$

The inequalities $g(x) \leqq 0$ and $g(x) < 0$ are taken component-wise.

We can now state the "standard" Karush-Kuhn-Tucker theorem, see, e.g., [11]. Note that a constraint qualification is a condition that guarantees the existence of a Kuhn-Tucker vector.

THEOREM 1 Suppose that some constraint qualification holds for (P), e.g.,

there exists $\hat{x} \in \Omega$ such that $g(\hat{x}) < 0$ (Slater's condition)

Then

$$\mu = \inf\{p(x) + \lambda g(x) : x \in \Omega\} \tag{4}$$

for some Kuhn-Tucker vector $\lambda = (\lambda_k) \geqq 0$. Moreover, if $\mu = p(x^*)$ for some x^* in F, then

$$\lambda g(x^*) = 0 \qquad \text{(complementary slackness)} \tag{5}$$

and (4) and (5) characterize optimality of x^* in F.

The inner product $\lambda g(x)$ of the two vectors λ and $g(x)$ is stated by juxtaposition.

Now if $f : U \to R$ is a convex function on U and $x \in U$, let

$$\partial f(x) = \{\phi \in R^n : \phi(y - x) \leqq f(y) - f(x), \text{ for all } y \text{ in } U\} \tag{6}$$

denote the *subdifferential of* f *at* x. For $K \subset R^n$, let

$$K^+ = \{\phi \in R^n : \phi y \geqq 0, \text{ for all } y \text{ in } K\} \tag{7}$$

be the *nonnegative polar cone* of K. Then we have the following optimality characterization for a convex function f on U.

LEMMA 1 (e.g., [10, p. 87]). Suppose that $x^* \in \Omega$. Then

$$f(x^*) = \inf\{f(x) : x \in \Omega\}$$

if and only if

$$\partial f(x^*) \cap (\Omega - x^*)^+ \neq \phi \tag{8}$$

If some constraint qualification holds for (P), this result and Theorem 1 yield:

$$\begin{aligned} &x^* \text{ in } F \text{ is optimal for (P) if and only if} \\ &0 \in \partial p(x^*) + \partial \lambda g(x^*) - (\Omega - x^*)^+ \\ &\text{for some } \lambda = (\lambda_k) \geqq 0 \text{ with } \lambda g(x^*) = 0 \end{aligned} \tag{9}$$

Note that p and λg are continuous on U, since they are convex and finite on the open set U. Thus

$$\partial(p(x^*) + \lambda g(x^*)) = \partial p(x^*) + \partial \lambda g(x^*)$$

The *directional derivative* of the convex function f at x in U in the direction d is

$$\nabla f(x;d) = \lim_{t \downarrow 0} \frac{f(x+td) - f(x)}{t} \tag{10}$$

Let

$$\mathcal{P}(x) = \{k \in \mathcal{P} : g^k(x) = 0\} \tag{11}$$

be the set of binding constraints at x in F. The set of *implicit equality constraints*, e.g., [1], is

$$\mathcal{P}^= = \{k \in \mathcal{P} : g^k(x) = 0, \text{ for all } x \text{ in } F\} \tag{12}$$

We set

$$\mathcal{P}^<(x) = \mathcal{P}(x) \backslash \mathcal{P}^= \tag{13}$$

$$C(x) = \{d \in R^n : \nabla g^k(x;d) \leqq 0, \text{ for all } k \in \mathcal{P}(x)\} \tag{14}$$

is the *linearizing cone* at x, while

$$B(x) = \text{cone} \bigcup_{k \in \mathcal{P}(x)} \partial g^k(x)$$

is the *cone of subgradients* at x, where cone denotes the generated convex cone.

$$T(F,x) = \overline{\text{cone}}(F - x) \tag{15}$$

is the tangent cone of the convex feasible set F at x, where $\overline{\cdot}$ denotes closure.

For the *perturbed program*

$$\mu(\varepsilon) = \inf\{p(x) : g(x) \leqq \varepsilon, \; x \in \Omega\} \qquad (P_\varepsilon)$$

$\mu(\varepsilon)$ is called the *perturbation function*. We let $F_\varepsilon = \{x : g(x) \leqq \varepsilon, \; x \in \Omega\}$ denote the *perturbed feasible set*.

The *cone of directions of constancy* of a convex function f at x is (see, e.g., [2])

$$D_f^{=}(x) = \{d \in R^n : \text{there exists } \bar{\alpha} > 0 \text{ with } f(x + \alpha d) = f(x), \text{ for all } 0 < \alpha \leqq \bar{\alpha}\} \qquad (16)$$

We let $D_k^{=}(x) = D_{g^k}^{=}(x)$ and $D_K^{=}(x) = \bigcap_{k \in K} D_k^{=}(x)$, where $K \subset P$. The *cones of directions of increase, decrease, and nonincrease*, denoted $D_f^{>}(x)$, $D_f^{<}(x)$, and $D_f^{\leqq}(x)$, are similarly defined. We also let

$$D_f^a(x) = \{d \in R^n : \text{there exists } \bar{\alpha} > 0, \text{ with } f(x + \alpha d) \text{ affine}, \text{ for all } 0 < \alpha \leqq \bar{\alpha}\}$$

denote the *cone of affine directions*; see [16].

III. RESTRICTED KUHN-TUCKER VECTORS

In this section we present several optimality conditions for (P) of the type (1), with Ω replaced by $\Omega \cap G$. These results extend the conditions given in [1,2,3,4] and [13,14,15,16] in the sense that the infimum need not be attained and Ω may be a proper subset of R^n. The proofs given here are unified and are based on Lemma 2 below. The idea is to find a regularizing set G, in order to satisfy the generalized Slater condition given in the lemma.

For each choice of the set G and corresponding vector λ which satisfies (1), with Ω replaced by $\Omega \cap G$, we find a set of stable perturbations; see Theorem 3 below.

We call $\lambda = (\lambda_k) \geqq 0$ a *restricted Kuhn-Tucker vector* with respect to the convex set G if

$$\mu \leqq \inf\{p(x) + \lambda g(x) : x \in \Omega_G = \Omega \cap G\} \qquad (17)$$

If the infimum for (P) is attained at x* in F, i.e., $\mu = p(x^*)$, then we can trivially choose $\lambda = 0$ and $G = \{x^*\}$. Or, in the case that the infimum is not attained, we can choose $\lambda = 0$ and $G = F$, the feasible set. Now consider the perturbation $\varepsilon = (\varepsilon_k)$ and suppose that (17) holds and

$$F_\varepsilon \subset \Omega_G \tag{18}$$

Then by (17),

$$\begin{aligned} p(x) - \mu &\geqq -\lambda g(x) && \text{for all } x \text{ in } \Omega_G \\ &\geqq -\lambda\varepsilon && \text{for all } x \text{ in } F_\varepsilon \text{ by (18)} \end{aligned}$$

Taking the infimum over x yields

$$\mu(\varepsilon) - \mu \geqq -\lambda\varepsilon \qquad \text{if (18) holds} \tag{19}$$

Thus the restricted Kuhn-Tucker vector λ provides a lower bound for the marginal improvement of the optimal value with respect to the perturbation ε if $F_\varepsilon \subset \Omega_G$. Thus the program is stable, or equivalently λ is a Kuhn-Tucker vector, if we can choose $G \supset \Omega$. Moreover, the program will be stable with respect to a larger set of perturbations ε if we can choose the set G larger. This motivates the search for larger sets G in (17) (or stronger optimality conditions, e.g., Gould and Tolle [8] and Guignard [9]). We will now present several choices for the set G in (17). First we need the following lemma. Note that a *polyhedral* function is the maximum of a finite number of affine functions.

LEMMA 2 [6] Consider the ordinary convex program (P). Suppose that Ω is a polyhedral set and that there exists an $x_1 \in \Omega$ such that $g^k(x_1) \leqq 0$, $k = 1,\ldots,m$, with strict inequality if g^k is not polyhedral (on Ω). Then

$$\mu = \inf\{p(x) + \lambda g(x) : x \in \Omega\} \tag{20}$$

for some $\lambda = (\lambda_k) \geqq 0$. Moreover, if $\mu = p(a)$ for some $a \in F$, then

$$\lambda g(a) = 0 \tag{21}$$

and (20) and (21) characterize optimality of a.

REMARK 1. If Ω is not polyhedral, then the above result holds if either $x_1 \in \mathrm{ri}\Omega$, the relative interior of Ω, or if the strict inequality, $g^k(x_1) < 0$, holds for all k in $\mathcal{P}$; see, e.g., [11].

Following Rockafellar [12], we say that a function $f : U \to R$ is *faithfully convex* (on U) if f is convex (on U) and f is affine on a line segment (in U) only if it is affine on the whole line (in U) containing that segment. Analytic convex and strictly convex functions are two examples of faithfully convex functions. If f is faithfully convex on U, then it can be shown that the cone of directions of constancy $D_f^{=}(x)$ is a subspace independent of the point x in U (see, e.g., [2]). We can now derive various choices for the set G in (17).

THEOREM 2 Suppose that Ω is polyhedral, g^k is faithfully convex (on U) for all $k \in P^{=}$, and $h = \sum_{k \in P^{=}} \alpha_k g^k$, where $\alpha_k \geqq 0$ with $\alpha_k > 0$ if g^k is not affine (on U). Let $\hat{x} \in F$. Then there exist convex sets G such that

$$\mu = \inf\{p(x) + \lambda g(x) : x \in \Omega_G = \Omega \cap G\} \tag{22}$$

for some $\lambda \geqq 0$ and, if $\mu = p(x^*)$ for some x^* in F, then

$$\lambda g(x^*) = 0 \tag{23}$$

and (22) and (23) characterize optimality of x^*. Possible choices of the convex set G in (22) are:

(a) $G_1 = \{x : g^k(x) = 0 \text{ for all } k \in P^{=}\}$

(b) $G_2 = \hat{x} + D_f^{=}(\hat{x})$, where $f \triangleq \max_{k \in P^{=}} g^k$

(c) $G_3 = \hat{x} + D_{P^{=}}^{=}(\hat{x})$

(d) $G_4 = \hat{x} + D_h^{=}(\hat{x})$

(e) $G_5 = \hat{x} + D_{P^{=}}^{a}(\hat{x})$

(f) $G_6 = \hat{x} + D_h^{a}(\hat{x})$

(Note that we can allow λ_k to be arbitrary for all k such that $g^k(x) = 0$ on $\Omega \cap G_i$.)

PROOF. (a): It can be shown that $G_1 = G_3$; see, e.g., [2]. Now see the proof of (c) below. (b) and (c): First let us show that $G_2 = G_3$. That $D_{P^{=}}^{=}(x) \subset D_f^{=}(x)$ is clear. Now suppose that $d \in D_f^{=}(x) \setminus D_{P^{=}}^{=}(x)$. Then, by definition of f, $d \in D_{P^{=}}^{\leqq}(x) \cap D_k^{<}(x)$ for some $k \in P^{=}$. It can be shown (see [2]) that there exists

$$\hat{d} \in D_{P^{=}}^{=}(\hat{x}) \cap D_{P^{<}(\hat{x})}^{<}(\hat{x}) \tag{24}$$

(The proof of (24) remains the same even though $\Omega \neq R^n$.) Let

$$d_\lambda = \lambda d + (1 - \lambda)\hat{d} \tag{25}$$

Then for sufficiently small $\lambda > 0$, we see that

$$d_\lambda \in D_{P(\hat{x})}^{\leqq}(\hat{x}) \cap D_k^{<}(\hat{x}) \tag{26}$$

which contradicts the definition of $P^=$. Thus

$$D_f^=(\hat{x}) = D_{P^=}^=(\hat{x})$$

Now it is clear that $F \subset \hat{x} + D_{P^=}^=(\hat{x})$. Thus an equivalent program to (P) is

$$\mu = \inf\{p(x) : g(x) \leqq 0, \ x \in \Omega \cap G_3\} \tag{27}$$

It is now sufficient to show that we can satisfy the hypothesis of Lemma 2 with Ω replaced by $\Omega \cap G_3$. Now be definition of $P^=$, for each $k \in P \backslash P^=$ there exists $x^k \in F$ such that $g^k(x^k) < 0$. Let t equal the cardinality of $P \backslash P^=$. Then, by the convexity of F, we see that

$$x_1 = \sum_{k \in P \backslash P^=} \frac{1}{t} x^k \in F$$

and

$$g^k(x_1) < 0, \ k \in P \backslash P^=; \quad g^k(x_1) = 0, \ k \in P^=; \quad x_1 \in \Omega \tag{28}$$

Moreover, this implies that

$$x_1 - \hat{x} \in D_{P^=}^=(\hat{x}) \tag{29}$$

Thus x_1 satisfies the constraint qualification in Lemma 2 if we show that

$$g^k \text{ is affine on } G_3 \cap \Omega \qquad \text{for all } k \in P^= \tag{30}$$

In fact, by the faithfully convex assumption, it is clear that g^k is 0 on $G_3 \cap \Omega$, for all $k \in P^=$. This proves that G_3, and so also G_2, is a possible choice for G in (22). Note that $\Omega \cap G_2$ is polyhedral since Ω is, by assumption, and G_2 is an affine subspace, by faithful convexity.

Following the details of the proof above, we see that to show that G_i, $i = 4,5,6$, are possible choices for G in (22) we need only show that

$$g^k \text{ is affine on } \Omega \cap G_i \qquad \text{for all } k \in P^=, \ i = 4,5,6 \tag{31}$$

(d): Let $d \in D_h^=$, $k_0 \in P^=$, and suppose g^{k_0} is not affine (on Ω). Then $\alpha_{k_0} > 0$ and

$$-\alpha_{k_0} g^{k_0}(\hat{x} + \alpha d) = \sum_{k \in P^= \backslash \{k_0\}} \alpha_k g^k(\hat{x} + \alpha d) \tag{32}$$

Since a nonnegative linear combination of convex functions is convex, this shows that both $\pm g^{k_0}(\hat{x} + \alpha d)$ is a convex function of α. Thus g^k is in fact an affine function of α. This shows that g^k is affine on $\Omega \cap G_4$ as desired.

The result with G_5 and G_6 follows similarly by allowing an affine term on the right-hand side of (32).

REMARK 2. The characterization of optimality using G_1 holds without the polydedrality and faithful convexity assumptions; see [5]. This extends the characterization of optimality given in [1,2,4,17], which requires that the infimum for (P) be attained and that $\Omega = \mathbb{R}^n$. Using Lemma 1, we get

$$
\begin{aligned}
&x^* \text{ (in } F\text{) is optimal for (P) if and only if} \\
&0 \in \partial p(x^*) + \partial \lambda g(x^*) - (\Omega_{1_{G_1}} - x^*)^+ \\
&\text{for some } \lambda = (\lambda_k) \geqq 0 \text{ with } \lambda g(x^*) = 0
\end{aligned}
$$

It can be shown (see [5]) that

$$(\Omega_{1_{G_1}} - x^*)^+ = \left(D^{=}_{P^=}(x^*) \cap \operatorname{cone}(\Omega - x^*)\right)^+$$

and that λ can be chosen with $\lambda_k = 0$ if $k \in P^=$. If we in fact fix $\Omega = R^n$ and choose λ with $\lambda_k = 0$ for $k \in P^=$, then (32) becomes the result in [4]. Note that this result is a special case of the generalized Kuhn-Tucker conditions given by Guignard [9].

We can similarly recover the results obtained in [13,14,15,16]. Moreover, one can replace the assumption of faithful convexity (and affinity) by piecewise faithful convexity (resp. polyhedrality), i.e., that g^k, $k \in P^=$, is the maximum of a finite number of faithfully convex functions. For, working with the equivalent program to (P), obtained by replacing the piecewise faithfully convex functions by the faithfully convex functions from which the maximum is taken, gives rise to a minorant for the Lagrangian, i.e.,

$$
\begin{aligned}
\mu &= \inf\left\{p(x) + \sum_{k,t} \lambda_{k,t} g^{k,t}(x) \; : \; x \in \Omega \cap G\right\} \\
&\leqq \inf\{p(x) + \lambda g(x) \; : \; x \in \Omega \cap G\}
\end{aligned}
$$

if $\lambda = (\lambda_k)$, $\lambda_k = \sum_t \lambda_{k,t}$, and $g^k = \max_t g^{k,t}$.†

We now define the following class of "stable" perturbations in R^m, with respect to a convex set G in R^n,

†This argument is due to Professor Jon Borwein.

$$\varepsilon(G) = \left\{\varepsilon \in R^m : \mu(\varepsilon) = \inf\{p(x) : g(x) \leqq \varepsilon,\ x \in \Omega \cap G\}\right\} \tag{33}$$

i.e., this is the set of perturbations for which the perturbation function does not change when the (regularizing) set G is added. We now have the following property of a restricted Kuhn-Tucker vector in terms of equilibrium prices for the perturbation.

THEOREM 3 When μ is finite, then $\lambda^* = (\lambda_k^*) \geqq 0$ is a restricted Kuhn-Tucker vector, with respect to the convex set G, implies that at the unit price λ_k^* for a perturbation ε_k, no perturbation in $\varepsilon(G)$ is worth buying.

PROOF. That no perturbation in $\varepsilon(G)$ is worth buying is equivalent to the inequality

$$\mu(\varepsilon) + \lambda^*\varepsilon \geqq \mu \qquad \text{for all } \varepsilon \in \varepsilon(G) \tag{34}$$

For any perturbation ε, the minimum value in the perturbed program (P_ε) plus the cost of buying the perturbation ε, at the unit prices given by the vector $v = (v_k)$, is

$$\mu(\varepsilon) + v\varepsilon$$

Now, for any m-vector $\lambda \geqq 0$,

$$\inf\{\mu(\varepsilon) + \lambda\varepsilon : \varepsilon \in \varepsilon(G)\} = \inf\{p(x) + \lambda\varepsilon : \varepsilon \in \varepsilon(G),\ x \in F_\varepsilon\}$$

$$= \inf\{p(x) + \lambda\varepsilon : \varepsilon \in \varepsilon(G),\ x \in \Omega,\ g(x) \leqq \varepsilon\}$$

$$= \inf\{p(x) + \lambda\varepsilon : \varepsilon \in \varepsilon(G),\ x \in \Omega_G,\ g(x) \leqq \varepsilon\} \qquad \text{by definition of } \varepsilon(G)$$

$$= \inf\{p(x) + \lambda g(x) : \varepsilon \in \varepsilon(G),\ x \in \Omega_G,\ g(x) \leqq \varepsilon\} \tag{35}$$

[since $\lambda \geqq 0$ and $\varepsilon_1 \in \varepsilon(G)$, $\varepsilon_2 \leqq \varepsilon_1$ implies $\varepsilon_2 \in \varepsilon(G)$]

$$\geqq \inf\{p(x) + \lambda g(x) : x \in \Omega_G\}$$

$$\geqq \mu \qquad \text{if and only if (17) holds, i.e., if } \lambda = \lambda^*$$

An alternative proof of the above is

$$p(x) - \mu \geqq -\lambda^* g(x) \qquad \text{for all } x \text{ in } \Omega \cap G, \text{ by assumption}$$

$$\geqq -\lambda^*\varepsilon \qquad \text{if } x \in \Omega \cap G \text{ and } g(x) \leqq \varepsilon$$

Thus

$$p(x) - \mu \geqq -\lambda^*\varepsilon \qquad \text{for all } x \text{ in } F_\varepsilon \text{ if } \varepsilon \in \varepsilon(G)$$

Taking the infimum over x on both sides yields

$$\mu(\varepsilon) + \lambda^*\varepsilon \geqq \mu$$

Note that if some constraint qualification holds then λ is a Kuhn-Tucker vector if and only if, at the unit prices λ_k, no perturbation whatsoever is economical to buy; see [11, p. 277]. It is conjectured that "if and only if" holds in the above theorem as well. For this to hold, it remains to show that equality holds in (35). Otherwise, we can find a larger set of perturbations than $\varepsilon(G)$.

COROLLARY 1 Suppose that the perturbation ε satisfies

$$F_\varepsilon \subset \Omega \cap G \tag{36}$$

and λ is a restricted Kuhn-Tucker vector satisfying (17). Then ε is not worth buying at the unit prices given by $\lambda = (\lambda_k)$.

PROOF. It is clear that (36) implies $\varepsilon \in \varepsilon(G)$.

The corresponding *restricted Lagrangian dual program* with respect to the set G is

$$\nu = \sup_{\lambda \geqq 0} \inf_{x \in \Omega \cap G} \{p(x) + \lambda g(x)\} \tag{D_G}$$

Note that if λ is a solution of (17), then $\nu = \mu$ (no duality gap) and λ is also a solution of (D_G).

IV. PROGRAMS WITH ATTAINED INFIMUM

Let us now assume that x^* is in F, $\mu = p(x^*)$, and both p and g are continuous at x^*. We now look for optimality conditions of the type

$$\begin{aligned} &x^* \text{ in } F \text{ is optimal if and only if} \\ &0 \in \partial p(x^*) + \partial \lambda g(x^*) - (\Omega_G - x^*)^+ \qquad \text{for some } \lambda \geqq 0 \text{ with } \lambda g(x^*) = 0 \end{aligned} \tag{37}$$

where $\Omega_G = \Omega \cap G$. By applying Lemma 1, these conditions can be reformulated in the same form as the conditions given in Theorem 2.

We now present some choices of G in (37) different and possibly larger than those which can be obtained from Theorem 2. These sets G can be found by satisfying the equation (x^* is an optimal solution)

$$T^+(F, x^*) = -B(x^*) + (\Omega_G - x^*)^+$$

In the differentiable case, $B(x^*)$ can be replaced by $C(x^*)$. (See [8,9,13] for more details.) Note that larger sets G in (17), and so "stronger" optimality conditions, correspond to "smaller" sets $(\Omega_G - x^*)$ in (37).

Now let

$$P^b(x) = \left\{k \in P^= : C(x) \cap D_k^>(x) \setminus \overline{D^=_{P^=}} \neq \emptyset\right\} \tag{38}$$

This is the set of *badly behaved constraints* at x (see [13]), i.e., the set of constraints which create problems in the Kuhn-Tucker theory. It can be shown [13] that

$$P^b(x) = \emptyset \quad \text{and} \quad B(x) \text{ is closed}$$

is a weakest constraint qualification at x, i.e., is a necessary and sufficient condition for the Kuhn-Tucker theory to hold at x. We can introduce the objective function into this definition by setting

$$P_p^b(x) = \left\{k \in P^= : C(x) \cap D_k^>(x) \cap D_p^<(x) \neq \emptyset\right\} \tag{39}$$

For simplicity we assume that $\Omega = R^n$ and g^k, $k \in P^=$, is faithfully convex and differentiable. Let $x \in F$ and

$$P^b(x) \subset K \subset P^=$$

Then we can choose (see [14])

$$G = \left(D_K^=(x)\right)^+$$

in (37). This is also true with $P^b(x)$ replaced by $P_p^b(x)$; see [14]. It can also be shown that $D_K^=(x)$ can be replaced by $D_K^a(x)$, $D_f^=(x)$, or $D_f^a(x)$, where $f = \sum_{k \in K} \alpha_k g^k$ and α_k are any positive scalars.

Moreover, it was shown in [14] that, discounting redundancies in the constraints, the shadow price λ_k corresponding to $k \in P_f^b(x^*)$ is essentially infinite, i.e., for $k \in P_f^b(x^*)$, small amounts of the perturbation ε_k are economical to buy no matter how high the unit cost.

EXAMPLE 1. Consider the (linear) program

$$\begin{aligned}
\mu = \mathit{inf}\ p(x) &= cx = x_1 + 2x_2 + 3x_3 \\
\text{s.t.}\ g^1(x) &= a^1x - b_1 = x_1 + x_2 + x_3 - 1 \leqq 0 \\
g^2(x) &= a^2x - b_2 = -x_1 - x_2 - x_3 + 1 \leqq 0 \\
g^3(x) &= a^3x - b_3 = 2x_1 + 2x_2 - 1 \leqq 0 \\
g^4(x) &= a^4x - b_4 = -2x_1 - 2x_2 + 1 \leqq 0 \\
g^5(x) &= a^5x - b_5 = -x_2 - x_3 + 1 \leqq 0 \qquad x \geqq 0
\end{aligned} \tag{40}$$

Then $P = P^{=} = \{1,2,\ldots,5\}$; $F = \{x_1 = 0,\ x_2 = x_3 = 1/2\}$; $\mu = 5/2$; and $G_1 = G_2 = G_3 = F$. Suppose we choose G_1 in Theorem 2. Since the constraints g^i are all identically 0 on G_1, any $\lambda \geqq 0$ solves (17) with $G = G_1$. However, not every $\lambda \geqq 0$ can be interpreted as a shadow price for the original program (P), e.g., $\lambda = 0$ is clearly not a shadow price for the perturbation $\varepsilon_k = 1$, for all $k \in P$. In addition, the corresponding restricted dual is identical to the primal program (40). Thus we get no duality information when using the regularizing set G_1.

Now let $\alpha_k = 1$ and $h(x) = \sum_{k\in P^{=}} \alpha_k g^k(x) = -x_2 - x_3 + 1$. Then

$$G_4 = \{x : x_2 + x_3 = 1\}$$

The corresponding restricted dual is

$$\mu = \sup_{\lambda \geqq 0} \inf_{\substack{x \geqq 0 \\ x_2+x_3=1}} \left\{cx + \sum_{k=1}^{5} \lambda_k (a^k x - b_k)\right\}$$

$$= \sup_{\lambda \geqq 0} \inf_{\substack{x \geqq 0 \\ x_2+x_3=1}} \left\{cx + \sum_{k=1}^{4} \lambda_k (a^k x - b_k)\right\} \qquad \text{since } g^5 \equiv 0 \text{ on } G_4$$

$$= \sup_{\lambda \geqq 0} \inf_{\substack{x \geqq 0 \\ x_2+x_3=1}} \left\{\left(c + \sum_{k=1}^{4} \lambda_k a^k\right)x - \sum_{k=1}^{4} \lambda_k b_k\right\}$$

$$= \sup_{\substack{\lambda \geqq 0 \\ 1+\lambda_1-\lambda_2+2\lambda_3-2\lambda_4 \geqq 0}} \inf_{\substack{x \geqq 0 \\ x_2+x_3=1}} \left\{(2 + \lambda_1 - \lambda_2 + 2\lambda_3 - 2\lambda_4)x_2 + (3 + \lambda_1 - \lambda_2)x_3 - \sum_{k=1}^{4} \lambda_k b_k\right\}$$

[since the infimum is $-\infty$ if $1 + \lambda_1 - \lambda_2 + 2\lambda_3 - 2\lambda_4 < 0$]

$$= \sup_{1+\alpha+2\beta \geqq 0} \inf_{\substack{x \geqq 0 \\ x_2+x_3=1}} \left\{(2 + \alpha + 2\beta)x_2 + (3 + \alpha)x_3 - (\alpha + \beta)\right\}$$

$$= 2\,\frac{1}{2} \qquad \text{when } \beta = \frac{1}{2}$$

Thus $\alpha \geqq -2$ and $\beta = 1/2$, i.e.,

$$\lambda_1 - \lambda_2 \geqq -2; \quad \lambda_3 - \lambda_4 = \frac{1}{2}; \quad \lambda_i \geqq 0$$

are the solutions of the restricted dual program. The perturbations $\varepsilon = (\varepsilon_k)$ with $\varepsilon_4, \varepsilon_5 \geqq 0$ and $\varepsilon_1 = \varepsilon_2 = \varepsilon_3 = 0$ are in $\varepsilon(G_4)$, since G_4 equals the set of points x which satisfy the constraints g^1, g^2, g^5. In particular, we get that $\nabla\mu(0;e_3) \geqq -(1/2)$ and $\nabla\mu(0;e_4) \geqq 0$, where e_3 and e_4 are the third and fourth unit vectors, respectively, i.e., we get shadow prices for the third and fourth constraints.

Since the constraints are affine, we have $G_5 = G_6 = R^n$. The corresponding dual reduces to the usual linear programming dual,

$$\begin{cases} \mu = \sup \ -\lambda b \\ \text{s.t. } A^t\lambda - c^t \geqq 0 \\ \qquad \lambda \geqq 0 \end{cases}$$

where A is the matrix with rows a^i and $\cdot^t$ denotes transpose. Two basic optimal solutions are $(0,2,\frac{1}{2},0,1)$ and $(0,0,\frac{1}{2},0,3)$. This yields $\nabla\mu(0;e_3) \geqq -(1/2)$ and $\nabla\mu(0;e_4) \geqq 0$, as before.

The example above shows how using larger choices of G in (17) yields more dual information. This is the case for nonlinear problems as well, as the following example illustrates.

EXAMPLE 2. Consider the program

$$\begin{cases} \mu = \inf \ p(x) = x_2 \\ \text{s.t. } g^1(x) = x_1^2 + x_2^2 - 1 \leqq 0 \\ \qquad g^2(x) = -x_1 \qquad + 1 \leqq 0 \end{cases}$$

Then $P = P^= = \{1,2\}$; $F = \{x_1 = 1, x_2 = 0\} = x^*$; $\mu = 0$; and $G_1 = G_2 = \dots = G_6 = F$. Choosing G_1 in Theorem 2 yields no information. Thus even though one of the constraints is affine, we cannot obtain any shadow prices. In fact, $\nabla\mu(0;e_i) = -\infty$ for the unit vectors e_1, e_2. This points out the fact that affine constraints cannot always be ignored. Note that $P^b(x^*) = P_p^b(x^*) = \{1\}$ and $D_1^=(x^*) = D_1^a(x^*) = \{0\}$. Thus using only the badly behaved constraints does not help. However, suppose that we change the objective function to be $p(x) = x_1$. Then G_i, $i = 1,\dots,6$, is unchanged. But now $P_p^b(x^*) = \emptyset$. We can therefore set $G = 0$ and get

$$0 = \begin{pmatrix}1\\0\end{pmatrix} + \lambda_1\begin{pmatrix}2\\0\end{pmatrix} + \lambda_2\begin{pmatrix}-1\\0\end{pmatrix} \qquad \lambda_i \geqq 0$$

This yields $\nabla\mu'(0;e_1) \geqq 0$; $\nabla\mu'(0;e_2) \geqq -1$.

REFERENCES

1. R. A. Abrams and L. Kerzner. A simplified test for optimality, *Journal of Optimization Theory and Applications*, Vol. 25, pp. 161-170 (1978).

2. A. Ben-Israel, A. Ben-Tal, and S. Zlobec. *Optimality in Nonlinear Programming: A Feasible Directions Approach*. Wiley-Interscience, New York, 1981.

3. A. Ben-Israel, A. Ben-Tal, and S. Zlobec. Characterizations of optimality in convex programming without a constraint qualification, *Journal of Optimization Theory and Applications*, Vol. 20, pp. 417-437 (1976).

4. A. Ben-Tal and A. Ben-Israel. Characterizations of optimality in convex programming: the nondifferentiable case, *Applicable Analysis*. pp. 137-156 (1979).

5. J. Borwein and H. Wolkowicz. Characterizations of optimality for the abstract convex program with finite dimensional range, *Journal of the Australian Mathematical Society*, Vol. 30, pp. 390-411 (1981).

6. J. Borwein and H. Wolkowicz. Cone-convex programming: stability and affine constraint functions, in *Generalized Concavity in Optimization and Economics* (S. Schaible and W. T. Ziemba, eds.), pp. 379-397, Academic Press, New York, 1981.

7. A. M. Geoffrion. Duality in nonlinear programming: A simplified applications-oriented development, *SIAM Review*, Vol. 13, pp. 1-37 (1971).

8. F. J. Gould and J. W. Tolle. Geometry of optimality conditions and constraint qualification, *Mathematical Programming*, Vol. 2, pp. 1-18 (1972).

9. M. Guignard. Generalized Kuhn-Tucker conditions for mathematical programming problems in Banach space, *SIAM Journal of Control*, Vol. 7, pp. 232-241 (1969).

10. R. B. Holmes. *Geometric Functional Analysis and Its Applications*. Springer-Verlag, New York, 1975.

11. R. T. Rockafellar. *Convex Analysis*. Princeton University Press, 1970.

12. R. T. Rockafellar. Some convex programs whose duals are linearly constrained, in *Nonlinear Programming* (J. B. Rosen, O. L. Mangasarian, and K. Ritter, eds.), pp. 293-322, Academic Press, New York, 1970.

13. H. Wolkowicz. Geometry of optimality conditions and constraint qualifications: The convex case, *Mathematical Programming*, Vol. 19, pp. 32-60 (1980).

14. H. Wolkowicz. Shadow prices for an unstable convex program, *Utilitas Mathematica*, Vol. 18, pp. 119-139 (1980).

15. H. Wolkowicz. A strengthened test for optimality, *Journal of Optimization Theory and Applications*, Vol. 35, pp. 497-515 (1981).

16. H. Wolkowicz. Optimality and the cone of affine directions, Research Report, The University of Alberta (1980).

17. S. Zlobec and A. Ben-Israel. Duality in convex programming: A linearization approach, *Mathematische Operationsforsch. Statist. Ser. Opt.*, Vol. 10, pp. 171-178 (1979).

CHAPTER 5 OPTIMAL VALUE CONTINUITY AND DIFFERENTIAL STABILITY BOUNDS UNDER THE MANGASARIAN-FROMOVITZ CONSTRAINT QUALIFICATION*

ANTHONY V. FIACCO / The George Washington University, Washington, D.C.

ABSTRACT

The continuity of the optimal value function and bounds on its upper and lower Dini derivatives are obtained for a general class of nonlinear parametric programs, using elementary and constructive arguments. An implicit function theorem is applied to transform a general parametric mathematical program into a locally equivalent inequality constrained program, and upper and lower bounds on the optimal value function directional derivative limit quotient are shown to hold for this reduced program. These bounds are then shown to apply in programs having both inequality and equality constraints where a parameter may appear anywhere in the program. This paper draws on several preliminary results reported by Fiacco and Hutzler for the inequality constrained problem and provides a number of extensions and missing proofs. These results generalize those provided by Gauvin and Tolle for the right-hand-side perturbation problem. Gauvin and Dubeau have essentially

*Serial T-435, 1 October 1980, Contract DAAG 29-79-C-0062, U.S. Army Research Office. This document has been approved for public sale and release; its distribution is unlimited.
The findings in this report are not to be construed as an official Department of the Army position, unless so designated by other authorized documents.

simultaneously obtained the same bounds using a completely different method of proof. Finally, it is noted that the continuity and bounds results for the general parametric program are immediate implications of the Gauvin-Tolle right-hand-side parameter results applied to an equivalent right-hand-side problem of higher dimension.

I. INTRODUCTION

Gauvin and Tolle [10], not assuming convexity but limiting their analysis to nonlinear programs involving right-hand-side perturbations, proved the continuity of the optimal value function and provided sharp bounds on the optimal value function upper and lower Dini derivatives when the problem functions are once continuously differentiable, the constraint set is uniformly compact, and the Mangasarian-Fromovitz Constraint Qualification (MFCQ) [19] holds at prescribed solution points. The purpose of this paper is to refine and continue the preliminary but incomplete study initiated by Fiacco and Hutzler [6], that extends the Gauvin-Tolle results to the general inequality constrained mathematical program in which a parameter appears arbitrarily in the constraints and the objective function. We complete the extension to a general parametric inequality-equality problem whose functions are once continuously differentiable.

In an essentially simultaneous paper, Gauvin and Dubeau [9] have independently obtained the same results under the same assumptions invoked here, but relying on some rather sophisticated implicit function theorems derived from results in Hestenes [13]. Their generalization follows the approach of Auslender [1], who provided similar bounds under a modified form of MFCQ, assuming the equality constraints to be once continuously differentiable and requiring the other problem functions to be only locally Lipschitz. Our method of proof is felt to be simpler and more constructive, involving only elementary arguments and allowing for a clearer interpretation and computational exploitation of the results.

Sections II and III, through Theorem 5, and Section V are taken more or less intact from Fiacco and Hutzler [6] and are reported here for completeness. Theorem 7 and Corollary 1 were also obtained in [6]. The remaining results, Lemmas 3, 4, 5 and Theorems 6, 8, 9, and 10 extend the reduction approach introduced in [6] and complete the direct development of the theory for the general inequality-equality constrained parametric problem.

A simple immediate indirect proof of the results, following easily from the results of Gauvin and Tolle [10] by applying these to a right-hand-side reformulation of the general parametric problem according to a suggestion of Janin [16] and Rockafellar [20], is noted in Section VI.

The reader is referred to the survey by Fiacco and Hutzler [7] for many related results and to the recent paper by Rockafellar [20] for a concise technical overview and a rather complete theoretical generalization and improvement of these and most of the subderivative and directional derivative results reported to date. The reader interested in extensions to infinite-dimensional spaces is referred to the works of Dem'janov and Pevnyi [4], Gollan [11], Lempio and Maurer [17], and Levitin [18].

II. NOTATION AND DEFINITIONS

In this paper we shall be concerned with mathematical programs of the form:

$$\begin{aligned} &\min_x \ f(x,\varepsilon) \\ &\text{s.t. } g_i(x,\varepsilon) \geqq 0 \qquad (i = 1,\ldots,m) \qquad\qquad P(\varepsilon) \\ &\qquad h_j(x,\varepsilon) = 0 \qquad (j = 1,\ldots,p) \end{aligned}$$

where $x \in E^n$ is the vector of decision variables, ε is a parameter vector in E^k, and the functions f, g_i, and h_j are once continuously differentiable on $E^n \times E^k$. The feasible region of problem $P(\varepsilon)$ will be denoted $R(\varepsilon)$ and the set of solutions $S(\varepsilon)$. The m-vector whose components are $g_i(x,\varepsilon)$, i = 1,...,m, and the p-vector whose components are $h_j(x,\varepsilon)$, j = 1,...,p, will be denoted by $g(x,\varepsilon)$ and $h(x,\varepsilon)$, respectively.

Following usual conventions the gradient, with respect to x, of a once differentiable real-valued function $f : E^n \times E^k \to E^1$ is denoted $\nabla_x f(x,\varepsilon)$ and is taken to be the row vector $[\partial f(x,\varepsilon)/\partial x_1,\ldots,\partial f(x,\varepsilon)/\partial x_n]$. If $g(x,\varepsilon)$ is a vector-valued function, $g : E^n \times E^k \to E^m$, whose components $g_i(x,\varepsilon)$ are differentiable in x, then $\nabla_x g(x,\varepsilon)$ denotes the $m \times n$ Jacobian matrix of g whose ith row is given by $\nabla_x g_i(x,\varepsilon)$, i = 1,...,m. The transpose of the Jacobian $\nabla_x g(x,\varepsilon)$ will be denoted $\nabla_x' g(x,\varepsilon)$. Differentiation with respect to the vector ε is denoted in a completely analogous fashion. Transposition of vectors and matrices is denoted by a prime.

The Lagrangian for $P(\varepsilon)$ will be written

$$L(x,\mu,\omega,\varepsilon) = f(x,\varepsilon) - \sum_{i=1}^{m} \mu_i g_i(x,\varepsilon) + \sum_{j=1}^{p} \omega_j h_j(x,\varepsilon)$$

and the set of Kuhn-Tucker vectors corresponding to the decision vector x will be given by

$$K(x,\varepsilon) = \left\{(\mu,\omega) \in E^m \times E^p : \nabla_x L(x,\mu,\omega,\varepsilon) = 0,\ \mu_i \geqq 0,\ \mu_i g_i(x,\varepsilon) = 0,\ i = 1,\ldots,m\right\}$$

Writing a solution vector as a function of the parameter ε, the index set for inequality constraints which are binding at a solution $x(\varepsilon)$ is denoted by $B(\varepsilon) = \{i : g_i(x(\varepsilon),\varepsilon) = 0\}$. Finally, the optimal value function will be defined as

$$f^*(\varepsilon) = \inf\{f(x,\varepsilon) : x \in R(\varepsilon)\}$$

Throughout this paper we shall make use of the well known Mangasarian-Fromovitz Constraint Qualification (MFCQ), which holds at a point $x \in R(\varepsilon)$ if:

1. There exists a vector $\tilde{y} \in E^n$ such that

$$\nabla_x g_i(x,\varepsilon)\tilde{y} > 0 \qquad \text{for } i \text{ such that } g_i(x,\varepsilon) = 0 \text{ and} \tag{1}$$

$$\nabla_x h_j(x,\varepsilon)\tilde{y} = 0 \qquad \text{for } j = 1,\ldots,p; \text{ and} \tag{2}$$

2. the gradients $\nabla_x h_j(x,\varepsilon)$, $j = 1,\ldots,p$, are linearly independent.

We will have occasion to make use of various continuity properties for both real-valued functions and point-to-set maps. There are several related definitions of the indicated properties. The ones most suited to our purpose follow. The reader interested in more detail is referred to Berge [2] and Hogan [15].

DEFINITION 1 Let $\phi : X \to Y$ be a point-to-set mapping and let $\{\varepsilon_n\} \subset X$, with $\varepsilon_n \to \bar{\varepsilon}$ in X as $n \to \infty$.

1. ϕ is said to be open at a point $\bar{\varepsilon}$ of X if, for each $\bar{x} \in \phi(\bar{\varepsilon})$, there exists a value n_0 and a sequence $\{x_n\} \subset Y$ with $x_n \in \phi(\varepsilon_n)$ for $n \geqq n_0$ and $x_n \to \bar{x}$.
2. ϕ is said to be closed at a point $\bar{\varepsilon}$ of X if $x_n \in \phi(\varepsilon_n)$ and $x_n \to \bar{x}$ together imply that $\bar{x} \in \phi(\bar{\varepsilon})$.

DEFINITION 2 A point-to-set mapping $\phi : X \to Y$ is said to be uniformly compact near a point $\bar{\varepsilon}$ of X if the closure of the set $\bigcup_{\varepsilon \text{ in } N(\bar{\varepsilon})} \phi(\varepsilon)$ is compact for some neighborhood $N(\bar{\varepsilon})$ of $\bar{\varepsilon}$.

In Section III we apply a reduction of variables technique to $P(\varepsilon)$ which transforms that program to a locally equivalent program involving only inequality constraints. This approach simplifies the derivation of intermediate results which are needed to derive the bounds on the directional derivative limit quotients of $f^*(\varepsilon)$ given in Section IV. A demonstration of the results is provided in the example of Section V. Section VI concludes with a few remarks concerning related results.

III. REDUCTION OF VARIABLES

In $P(\varepsilon)$, if the rank of the Jacobian, $\nabla_x h$, with respect to x of the (first n) equality constraints in a neighborhood of a solution is equal to n, then the given solution is completely determined as a solution of the system of equations $h_j(x,\varepsilon) = 0$, $j = 1,\ldots,n$, and the (locally unique) solution, $x(\varepsilon)$, of this system near $\varepsilon = 0$ is then completely characterized by the usual implicit function theorem. We are here interested in the less structured situation and hence assume that the rank of $\nabla_x h$ is less than n. Since we shall be making use of MFCQ, this entails the assumption that the number p of equality constraints is less than n. If there are no equality constraints in a particular formulation of $P(\varepsilon)$, simply suppress reference to h in the following development. Otherwise, we take advantage of the linear independence assumption to eliminate the equalities, again using an implicit function theorem.

Let $x = (x_D, x_I)$, where $x_D \in E^p$ and $x_I \in E^{n-p}$. Reordering variables if necessary, if $h(x^*,\varepsilon^*) = 0$ and MFCQ holds at x^*, then we may assume that $\nabla_{x_D} h$ is nonsingular at (x^*,ε^*). Then the usual implicit function theorem results hold: there exists an open set $N^* \subset E^{n-p} \times E^k$ containing (x_I^*,ε^*) such that the system of equations $h(x_D,x_I,\varepsilon) = 0$ can be solved for x_D in terms of x_I and ε for any (x_I,ε) in N^*. Furthermore, this representation is unique, the resulting function $x_D = x_D(x_I,\varepsilon)$ is continuous, and $x_D^* = x_D(x_I^*,\varepsilon^*)$. Thus, in N^*, the system $h(x_D,x_I,\varepsilon) = 0$ is satisfied identically by the function $x_D = x_D(x_I,\varepsilon)$. Under our additional ssumption that h is once continuously differentiable in x_I and ε, $x_D(x_I,\varepsilon)$ is also once continuously differentiable in x_I and ε.

Applying this result to $P(\varepsilon)$ at x^*, since we have $\tilde{h}(x_I,\varepsilon) \equiv h[x_D(x_I,\varepsilon),x_I,\varepsilon] \equiv 0$ in N^*, this problem can be reduced locally to one involving only inequality constraints:

$$\min_{x_I} \tilde{f}(x_I,\varepsilon)$$
$$\text{s.t.} \quad \tilde{g}_i(x_I,\varepsilon) \geqq 0 \qquad (i = 1,\ldots,m) \qquad \tilde{P}(\varepsilon)$$
$$\text{and} \quad (x_I,\varepsilon) \in N^*$$

where $\tilde{f}(x_I,\varepsilon) \equiv f[x_D(x_I,\varepsilon),x_I,\varepsilon]$ and $\tilde{g}_i(x_I,\varepsilon) \equiv g_i[x_D(x_I,\varepsilon),x_I,\varepsilon]$ for $i = 1,\ldots,m$, and where the minimization is now performed over the $n - p$ dimensional vector x_I. The programs P and $\tilde{P}$ are locally equivalent, for (x,ε) in a neighborhood T^* of $(x^*,0)$ and for (x_I,ε) in N^*, in the sense that the point $x(\varepsilon) \in E^n$, with $(x(\varepsilon),\varepsilon)$ in T^* and $x(\varepsilon) = (x_D(\varepsilon),x_I(\varepsilon))$, satisfies the Karush-Kuhn-Tucker first-order necessary conditions for an optimum of $P(\varepsilon)$ if and only if the point $x_I(\varepsilon)$, with $(x_I(\varepsilon),\varepsilon) \in N^*$, satisfies those conditions for $\tilde{P}(\varepsilon)$, where $x_D(x_I,\varepsilon)$ is as given above. Furthermore, in the given neighborhoods, $x(\varepsilon) = (x_D(x_I(\varepsilon),\varepsilon),x_I(\varepsilon))$ is a local solution of $P(\varepsilon)$ if and only if $x_I(\varepsilon)$ is a local solution of $\tilde{P}(\varepsilon)$.

We first observe that the Mangasarian-Fromovitz Constraint Qualification for $P(\varepsilon)$ is inherited by the reduced problem $\tilde{P}(\varepsilon)$. For simplicity in notation, and without loss of generality, assume that $\varepsilon^* = 0$, and assume that the components of x have been relabeled so that $x = (x_D,x_I)$ and $\nabla_{x_D} h(x_D^*,x_I^*,0)$ is nonsingular. The next result is easily obtained by invoking MFCQ at $(x,\varepsilon) = (x^*,0)$, partitioning the MFCQ vector $\tilde{y} = (\tilde{y}_D,\tilde{y}_I)$ in conformance with $x^* = (x_D^*,x_I^*)$, differentiating $\tilde{h}$ and $\tilde{g}$ with respect to x_I, and applying the assumptions. Corresponding to the notation for $P(\varepsilon)$, we denote the feasible region, solution set, and optimal value of $\tilde{P}(\varepsilon)$ by $\tilde{R}(\varepsilon)$, $\tilde{S}(\varepsilon)$, and $\tilde{f}^*(\varepsilon)$, respectively. Other corresponding problem constituents will be similarly denoted.

LEMMA 1 If $g,h \in C^1$, then MFCQ holds at $x^* \in R(0)$, the feasible region of $P(0)$, with $\tilde{y} = (\tilde{y}_D,\tilde{y}_I) \in E^n$ the associated vector, where $\tilde{y}_D \in E^p$ and $\tilde{y}_I \in E^{n-p}$, if and only if MFCQ holds at the point $x_I^* \in \tilde{R}(0)$, the feasible region of $\tilde{P}(0)$, with vector $\tilde{y}_I$.

PROOF. Suppose that MFCQ holds for $P(0)$ at $(x^*,0) = (x_D^*,x_I^*,0)$ with $\tilde{y} = (\tilde{y}_D,\tilde{y}_I)$ the associated vector. Writing $\nabla_x h$ as $[\nabla_{x_D} h : \nabla_{x_I} h]$, we see that (2) can be expressed as:

$$\nabla_{x_D} h(x^*,0)\tilde{y}_D + \nabla_{x_I} h(x^*,0)\tilde{y}_I = 0 \tag{3}$$

Since we have assumed that $\nabla_{x_D} h(x^*,0)$ is nonsingular, we can solve for $\tilde{y}_D$ in (3) and obtain:

$$\tilde{y}_D = -\left[\nabla_{x_D} h(x^*,0)\right]^{-1} \nabla_{x_I} h(x^*,0)\tilde{y}_I \tag{4}$$

Now, denoting the inequality constraints of $\tilde{P}(0)$ by $\tilde{g}_i$, i.e., $\tilde{g}_i = g_i(x_D(x_I,0),x_I,0)$, $i = 1,\ldots,m$, by differentiating with respect to x_I we obtain:

$$\nabla_{x_I}\tilde{g}_i = \nabla_{x_D} g_i \; \nabla_{x_I} x_D + \nabla_{x_I} g_i$$

or

$$\nabla_{x_I}\tilde{g}_i = \nabla_x g_i \begin{bmatrix} \nabla_{x_I} x_D \\ \\ I \end{bmatrix} \tag{5}$$

where I is the identity matrix. Multiplying by $\tilde{y}_I$ in (5) we have:

$$\nabla_{x_I}\tilde{g}_i\tilde{y}_I = \nabla_x g_i \begin{bmatrix} \nabla_{x_I} x_D \\ \\ I \end{bmatrix} \tilde{y}_I \tag{6}$$

but $h(x_D(x_I,0),x_I,0) \equiv 0$, so that

$$\nabla_{x_D} h(x_D(x_I,0),x_I,0)\nabla_{x_I} x_D + \nabla_{x_I} h(x_D(x_I,0),x_I,0) = 0$$

and since $\nabla_{x_D} h(x_D(x_I,0),x_I,0)$ is nonsingular, we obtain:

$$\nabla_{x_I} x_D = -\left[\nabla_{x_D} h(x_D(x_I,0),x_I,0)\right]^{-1} \nabla_{x_I} h(x_D(x_I,0),x_I,0)$$

Substituting this last expression in (6) we have:

$$\nabla_{x_I}\tilde{g}_i\tilde{y}_I = \nabla_x g_i \begin{bmatrix} -\left[\nabla_{x_D} h(x_D(x_I,0),x_I,0)\right]^{-1}\nabla_{x_I} h(x_D(x_I,0),x_I,0) \\ \\ I \end{bmatrix} \tilde{y}_i$$

and from (4) we see that at $(x_I^*,0)$,

$$\nabla_{x_I}\tilde{g}_i\tilde{y}_I = \nabla_x g_i \begin{bmatrix} \tilde{y}_D \\ \tilde{y}_I \end{bmatrix} = \nabla_x g_i \tilde{y} \tag{7}$$

Thus, by (1) it follows that $\nabla_{x_I}\tilde{g}_i\tilde{y}_I > 0$.

In [10], Gauvin and Tolle established that the set of Kuhn-Tucker multipliers associated with a solution, x^*, of $P(0)$ is nonempty, compact, and convex if and only if MFCQ is satisfied at x^*. That result enables us to establish in Theorem 1 a necessary link between a directional derivative, with respect to the decision variable x_I, of the objective function at an optimal point and a directional derivative of the Lagrangian taken with respect to the parameter ε. It is this relationship that eventually leads to the upper and lower bounds on the directional derivative limit quotients, which are derived in the next section.

We now obtain several perturbation results for problem $\tilde{P}(\varepsilon)$. These *do not* depend on the variable-reduction derivation of $\tilde{P}(\varepsilon)$ and are applicable to any inequality constrained problem having the indicated structure. Hence, unless otherwise stipulated, we assume in the following that a problem of form $\tilde{P}(\varepsilon)$ is given, without reference to $P(\varepsilon)$. It is further assumed that N^* is open and that $\tilde{f}$ and the $\tilde{g}_i$ are once continuously differentiable on N^*.

The next two theorems are crucial in obtaining the sharp bounds on the optimal value directional derivative limit quotient. They show that, at a local minimum where MFCQ holds, there exists a direction (in E^{n-p}) in which the directional derivative of the objective function yields that portion of the bound attributable to the constraint perturbation. For simplicity, the subscript x_I will be dropped from the gradient ∇ taken with respect to x_I.

THEOREM 1 If the conditions of MFCQ are satisfied for some $\bar{x}_I \in \tilde{S}(0)$, then, for any direction $z \in E^k$, there exists a vector $\bar{y} \in E^{n-p}$ satisfying:

1. $-\nabla\tilde{g}_i(\bar{x}_I,0)\bar{y} \leqq \nabla_\varepsilon\tilde{g}_i(\bar{x}_I,0)z$ for $i \in \tilde{B}(0)$, and (8)
2. $\nabla\tilde{f}(\bar{x}_I,0)\bar{y} = \max_{\mu\in\tilde{K}(\bar{x}_I,0)} [-\mu'\nabla_\varepsilon\tilde{g}(\bar{x}_I,0)z]$ (9)

PROOF. Given $z \in E^k$, consider the following linear program:

$$\max_\mu \; [-\mu'\nabla_\varepsilon\tilde{g}(\bar{x}_I,0)z]$$
$$\text{s.t. } \mu'\nabla\tilde{g}(\bar{x}_I,0) = \nabla\tilde{f}(\bar{x}_I,0)$$
$$\mu_i\tilde{g}_i(\bar{x}_I,0) = 0 \qquad (i = 1,\ldots,m)$$
$$\mu_i \geqq 0 \qquad (i = 1,\ldots,m)$$

The dual of this program is given by:

$$\min_{y,v} \quad \nabla\tilde{f}(\bar{x}_I,0)y$$
$$\text{s.t.} \quad \nabla\tilde{g}_i(\bar{x}_I,0)y + \tilde{g}_i(\bar{x}_I,0)v_i \geqq -\nabla_\varepsilon g_i(\bar{x}_I,0)z \qquad (i = 1,\ldots,m)$$
$$y \in E^{n-p} \qquad\qquad v_i \text{ unrestricted}$$

Since MFCQ is assumed to hold at $(\bar{x}_I,0)$, from [10] we have that $\tilde{K}(\bar{x}_I,0)$ is nonempty, compact, and convex. Thus, the primal problem is bounded and feasible. By the duality theorem of linear programming, the dual program has a solution, $(\bar{y},\bar{v})$, and hence there exists a vector $\bar{y}$ satisfying (8) and (9).

In the next two theorems we show first that, along any ray emanating from $\varepsilon = 0$, $\tilde{P}(\varepsilon)$ has points of feasibility near $\varepsilon = 0$, and second, that the existence of feasible points is guaranteed not only along rays but in a full neighborhood of $\varepsilon = 0$. In obtaining the following results associated with problem $\tilde{P}(\varepsilon)$, it is assumed that the analysis is confined to (x_I,ε) in N*.

THEOREM 2 If MFCQ holds at $\bar{x}_I \in \tilde{S}(0)$ then, for any unit vector $z \in E^k$ and any $\delta > 0$, $\tilde{g}(\bar{x}_I + \beta(\bar{y} + \delta\tilde{y}_I),\beta z) > 0$ for β positive and sufficiently near zero, where $\bar{y}$ is any vector satisfying (8) and $\tilde{y}_I$ satisfies MFCQ.

PROOF. Let z be any unit vector in E^k and consider first the case in which the constraint $\tilde{g}_i(x_I,\varepsilon) \geqq 0$ is binding at $(\bar{x}_I,0)$. Expanding $\tilde{g}_i(\bar{x}_I + \beta(\bar{y} + \delta\tilde{y}_I),\beta z)$ about the point $(\bar{x}_I,0)$, we obtain:

$$\begin{aligned}\tilde{g}_i(\bar{x}_I + \beta(\bar{y} + \delta\tilde{y}_I),\beta z) &= \beta\nabla_x\tilde{g}_i(\bar{x}_I + t\beta(\bar{y} + \delta\tilde{y}_I),\beta z)(\bar{y} + \delta\tilde{y}_I) \\ &\quad + \beta\nabla_\varepsilon\tilde{g}_i(\bar{x}_I,t'\beta z)z \\ &= \beta\left[\nabla_x\tilde{g}_i(\bar{x}_I + t\beta(\bar{y} + \delta\tilde{y}_I),\beta z)\bar{y} + \nabla_\varepsilon\tilde{g}_i(\bar{x}_I,t'\beta z)z\right] \\ &\quad + \beta\delta\nabla_x\tilde{g}_i(\bar{x}_I + t\beta(\bar{y} + \delta\tilde{y}_I),\beta z)\tilde{y}_I\end{aligned}$$

where $t,t' \in (0,1)$ and $t = t(\beta)$, $t' = t'(\beta)$.

Now, by (1), $\nabla_x\tilde{g}_i(\bar{x}_I,0)\tilde{y}_I = a_i > 0$. Thus, there exists $\beta' > 0$ such that for all $\beta \in [0,\beta']$,

$$\nabla_x\tilde{g}_i(\bar{x}_I + t\beta(\bar{y} + \delta\tilde{y}_I),\beta z)\tilde{y}_I \geqq (3a_i/4)$$

From (8) it follows that for β sufficiently small,

$$\nabla_x\tilde{g}_i(\bar{x}_I + t\beta(\bar{y} + \delta\tilde{y}_I),\beta z)\bar{y} + \nabla_\varepsilon\tilde{g}_i(\bar{x}_I,t'\beta z)z \geqq -(\delta a_i/4)$$

Thus, for β positive and near zero we have:

$$\tilde{g}_i(\bar{x}_I + \beta(\bar{y} + \delta\tilde{y}_I), \beta z) \geqq \beta(-(\delta a_i/4)) + \beta\delta(3a_i/4) = (\beta\delta a_i/2) > 0$$

Finally, if $\tilde{g}_i(\bar{x}_I, 0) > 0$, since each $\tilde{g}_i$ is jointly continuous in x and ε, it follows that, for any unit vector $z \in E^k$, and any $\delta > 0$, $\tilde{g}_i(\bar{x}_I + \beta(\bar{y} + \delta\tilde{y}_I), \beta z) > 0$ for β near zero.

THEOREM 3 If MFCQ is satisfied at $\bar{x}_I \in \tilde{S}(0)$ and if $\varepsilon_k \to 0$, then, given any $\delta > 0$, there exists $\beta_{k_j} \to 0$ and a vector $\bar{y}$ such that $\tilde{g}(\bar{x}_I + \beta_{k_j}(\bar{y} + \delta\tilde{y}_I), \varepsilon_{k_j}) \geqq 0$ for large j, where $\{\varepsilon_{k_j}\} \subseteq \{\varepsilon_k\}$, $\varepsilon_{k_j} \equiv ||\varepsilon_{k_j}|| z_{k_j}$, $z_{k_j} \to \bar{z}$, $\bar{y}$ satisfies (8) for $z = \bar{z}$, and $\tilde{y}_I$ is given by MFCQ.

PROOF. If $\{\varepsilon_{k_j}\} \subseteq \{\varepsilon_k\}$ and $\varepsilon_{k_j} \equiv 0$ for every j, the conclusion follows for $\{\varepsilon_{k_j}\}$ and any $\bar{y}$ by taking $\beta_{k_j} = 0$ for every j. Suppose $\varepsilon_{k_j} \neq 0$ for every j. Define $\beta_{k_j} = ||\varepsilon_{k_j}||$ and $z_{k_j} = \varepsilon_{k_j} / ||\varepsilon_{k_j}||$. Then, relabeling z_{k_j} if necessary, we can assume there exists $\bar{z}$ such that $z_{k_j} \to \bar{z}$. Let $\bar{y}$ satisfy (8) for the vector $\bar{z}$. Then, from Theorem 2 and the continuity of $\tilde{g}$ it follows that $\tilde{g}(x_I + \beta_{k_j}(\bar{y} + \delta\tilde{y}_I), \varepsilon_{k_j}) \equiv \tilde{g}(\bar{x}_I + \beta_{k_j}(\bar{y} + \delta\tilde{y}_I), \beta_{k_j} z_{k_j}) > 0$ for large j.

By Theorem 3, the satisfaction of the Mangasarian-Fromovitz Constraint Qualification at a solution point $\bar{x}_I$ of $\tilde{P}(0)$ is enough to guarantee the existence of feasible points for $\tilde{P}(\varepsilon)$ near $\bar{x}_I$. One might suspect that there exist points feasible to $\tilde{P}(\varepsilon)$ that are also feasible to $\tilde{P}(0)$. This is indeed the case, as the next theorem implies (see the statement immediately following the proof of Theorem 4). We shall need Theorem 4 in obtaining one of the key results in Theorem 8.

THEOREM 4 Let $\beta_n \to 0^+$ in E^1, let z be any unit vector in E^k, and let $\delta > 0$. If $x_I^n \in \tilde{R}(\beta_n z)$, with $x_I^n \to \bar{x}_I \in \tilde{R}(0)$, and if the conditions of MFCQ are satisfied at $\bar{x}_I$, then $x_I^n + \beta_n(\bar{y} + \delta\tilde{y}_I) \in \tilde{R}(0)$ for n sufficiently large, where $\bar{y}$ satisfies (8) with z replaced by -z, and $\tilde{y}_I$ is given by the constraint qualification.

PROOF. Consider first the case that $i \in \tilde{B}(0)$. Expanding $\tilde{g}_i(x_I^n + \beta_n(\bar{y} + \delta\tilde{y}_I), 0)$ about the point $(x_I^n, \beta_n z)$, we obtain:

$$\begin{aligned}\tilde{g}_i(x_I^n + \beta_n(\bar{y} + \delta\tilde{y}_I), 0) &= \tilde{g}_i(x_I^n, \beta_n z) \\ &\quad + \beta_n \nabla\tilde{g}_i(x_I^n + t\beta_n(\bar{y} + \delta\tilde{y}_I), 0)(\bar{y} + \delta\tilde{y}_I) \\ &\quad - \beta_n \nabla_\varepsilon \tilde{g}_i(x_I^n + \beta_n(\bar{y} + \delta\tilde{y}_I), t'\beta_n z)z\end{aligned}$$

where $t, t' \in (0,1)$, $t = t(\beta_n)$, $t' = t'(\beta_n)$. If, for n large, $x_I^n + \beta_n(\bar{y} + \delta\tilde{y}_I) \notin \tilde{R}(0)$, since x_I^n is feasible for $\tilde{P}(\beta_n z)$, it must be that

$$\beta_n \nabla\tilde{g}_i(x_I^n + t\beta_n(\bar{y} + \delta\tilde{y}_I),0)(\bar{y} + \delta\tilde{y}_I) < \beta_n \nabla_\varepsilon \tilde{g}_i(x_I^n + \beta_n(\bar{y} + \delta\tilde{y}_I),t'\beta_n z)z \tag{10}$$

Dividing by β_n in (10) and taking the limit as $n \to \infty$, we have

$$\nabla\tilde{g}_i(\bar{x}_I,0)(\bar{y} + \delta\tilde{y}_I) < \nabla_\varepsilon \tilde{g}_i(\bar{x}_I,0)z$$

But this contradicts (8) with z replaced by -z in (8), since $\delta > 0$ and by MFCQ $\nabla\tilde{g}_i(\bar{x}_I,0)\tilde{y}_I > 0$.

If, on the other hand, $i \notin \tilde{B}(0)$, $\tilde{g}_i(x_I^n + \beta_n(\bar{y} + \delta\tilde{y}_I),0) > 0$ for large n by the continuity of $\tilde{g}_i$ and the fact that $x_I^n \to \bar{x}_I$ and $\beta_n \to 0$.

It may be interesting to note, that by taking $x_I^n = \bar{x}_I + \beta_n(\bar{y} + \delta\tilde{y}_I)$ for each n in the hypothesis of Theorem 4, then Theorems 3 and 4 together imply that $\tilde{R}(\varepsilon)$ and $\tilde{R}(0)$ have points in common for ε near 0.

Let M be a closed subset of N*, selected to satisfy additional properties as needed in the following development, and denote by $\bar{P}(\varepsilon)$ the problem $\tilde{P}(\varepsilon)$, where N* is replaced by M. The feasible region of $\bar{P}(\varepsilon)$ will be denoted by $\bar{R}(\varepsilon)$, the solution set by $\bar{S}(\varepsilon)$, the optimal value by $\bar{f}^*(\varepsilon)$, etc. Denote by M^0 the interior of M.

We now show that the optimal value functions $\bar{f}^*(\varepsilon)$ and $f^*(\varepsilon)$ of $\bar{P}(\varepsilon)$ and $P(\varepsilon)$, respectively, are continuous near $\varepsilon = 0$ under the given assumptions. This result was proved by the author under slightly more general assumptions [5], and will be needed in the proof of Theorem 8. The continuity of $f^*(\varepsilon)$ was also shown by Gauvin and Dubeau [9], under the same assumptions as those given here for $P(\varepsilon)$.

LEMMA 2 $\bar{R}(\varepsilon)$ is a closed mapping at $\varepsilon = 0$.

PROOF. This follows immediately from the continuity of the $\tilde{g}_i$ and the fact that M is closed.

THEOREM 5 If $\bar{R}(\varepsilon)$ is uniformly compact for ε near zero and if there exists $\bar{x}_I \in \bar{S}(0)$ such that $(\bar{x}_I,0) \in M^0$ and the conditions of MFCQ hold at $\bar{x}_I$, then $\bar{f}^*(\varepsilon)$ is continuous at $\varepsilon = 0$.

PROOF. Let $\varepsilon_n \to 0$ in E^k be such that $\underline{\lim}_{\varepsilon\to 0} \bar{f}^*(\varepsilon) = \underline{\lim}_{n\to\infty} \bar{f}^*(\varepsilon_n)$. Clearly, since $\bar{R}(\varepsilon_n) \neq \phi$ for n sufficiently large (Theorem 3), $\bar{S}(\varepsilon_n) \neq \phi$ for large n. Hence, assuming n is large enough, there exists $x_I^n \in \bar{S}(\varepsilon_n)$. By the uniform compactness of $\bar{R}(\varepsilon)$, the sequence $\{x_I^n\}$ admits a convergent subsequence $\{x_I^{n_j}\}$. Let $\bar{\bar{x}}_I$ denote the limit of that subsequence. From Lemma 2, $\bar{R}(\varepsilon)$ is a closed mapping at $\varepsilon = 0$, so $\bar{\bar{x}}_I \in \bar{R}(0)$. Thus,

$$\underline{\lim_{\varepsilon\to 0}}\ \bar{f}^*(\varepsilon) = \lim_{j\to\infty} \bar{f}^*(\varepsilon_{n_j}) = \lim_{j\to\infty}\left(\tilde{f}\ x_I^{n_j},\varepsilon_{n_j}\right) = \tilde{f}(\bar{\bar{x}}_I,0) \geqq \bar{f}^*(0)$$

and we see that $\bar{f}^*(\varepsilon)$ is lsc at $\varepsilon = 0$.

Now let $\delta > 0$, let $\tilde{y}_I$ be given by MFCQ for $\bar{x}_I$, and select $\varepsilon_n \to 0$ such that $\overline{\lim_{\varepsilon\to 0}}\ f^*(\varepsilon) = \overline{\lim_{n\to\infty}}\ \bar{f}^*(\varepsilon_n)$. From Theorem 3 we know there exists $\beta_{n_j} \to 0$ and $\bar{y}$ such that $\bar{x}_I + \beta_{n_j}(\bar{y} + \delta\tilde{y}_I) \in \tilde{R}(\varepsilon_{n_j})$ for j large, where $\bar{y}$ satisfies (8) for some vector z. Hence

$$\overline{\lim_{\varepsilon\to 0}}\ f^*(\varepsilon) = \overline{\lim_{j\to\infty}}\ \bar{f}^*(\varepsilon_{n_j}) \leqq \overline{\lim_{j\to\infty}}\ \bar{f}\left(\bar{x}_I + \beta_{n_j}(\bar{y} + \delta\tilde{y}_I),\varepsilon_{n_j}\right) = \bar{f}(\bar{x}_I,0)$$
$$= \bar{f}^*(0)$$

Thus $\bar{f}^*(\varepsilon)$ is also usc at $\varepsilon = 0$ and we may conclude that $\bar{f}^*(\varepsilon)$ is continuous at $\varepsilon = 0$.

We should mention that the continuity of f^* requires only the *continuity* of f, in addition to the once (joint) continuous differentiability of the constraints.

The continuity of $\bar{f}^*(\varepsilon)$ at $\varepsilon = 0$ leads to a simple proof of the continuity of $f^*(\varepsilon)$, the optimal value of $P(\varepsilon)$. This is of intrinsic interest and will also be used in deriving the directional derivative limit quotient lower bound in the sequel. We first note the following result, the first part being an easy consequence of the continuity of the problem functions. The proof that $f^*(\varepsilon)$ is lsc at $\varepsilon = 0$ precisely parallels the first part of the proof of Theorem 5 that shows that $\bar{f}^*(\varepsilon)$ is lsc.

LEMMA 3 If $R(0)$ is nonempty and $R(\varepsilon)$ is uniformly compact for ε near 0, then $R(\varepsilon)$ is a closed mapping and $f^*(\varepsilon)$ is lsc at $\varepsilon = 0$.

THEOREM 6 If $R(0)$ is nonempty, $R(\varepsilon)$ is uniformly compact for ε near 0, and if MFCQ holds at some $x^* \in S(0)$, then $f^*(\varepsilon)$ is continuous at $\varepsilon = 0$.

PROOF. We eliminate the equalities of $P(\varepsilon)$ at x^* for (x_I,ε) in a neighborhood N^* of $(x_I^*,0)$, where $x^* = (x_D^*,x_I^*)$, using the previously defined variable reduction transformation, constructing problems of the form $\tilde{P}(\varepsilon)$ and $\bar{P}(\varepsilon)$, where the closed subset M of N^* is selected so that $(x_I^*,0) \in M^0$. We know that $x_I^* \in \tilde{S}(0)$ and that MFCQ holds at x_I^* (Lemma 1). Also, the uniform compactness of $R(\varepsilon)$ near $\varepsilon = 0$ implies the uniform compactness of $\bar{R}(\varepsilon)$ near $\varepsilon = 0$.

Clearly, $f^*(\varepsilon) \leqq \tilde{f}^*(\varepsilon) \leqq \bar{f}^*(\varepsilon)$, and since $f^*(0) = f(x^*,0) = \tilde{f}(x_I^*,0)$, we conclude that $\bar{f}^*(0) = \tilde{f}^*(0) = f^*(0)$, which also implies that $x_I^* \in \bar{S}(0)$. The assumptions of Theorem 5 are satisfied, hence $\bar{f}^*(\varepsilon)$ is continuous at 0. These relationships imply that $\overline{\lim}_{\varepsilon\to 0} f^*(\varepsilon) \leqq \overline{\lim}_{\varepsilon\to 0} \bar{f}^*(\varepsilon) = \bar{f}^*(0) = f^*(0)$; i.e., $f^*(\varepsilon)$ is usc at 0. Since $f^*(\varepsilon)$ is also lsc at $\varepsilon = 0$ (Lemma 3), the conclusion follows.

IV. BOUNDS ON THE PARAMETRIC VARIATION OF THE OPTIMAL VALUE FUNCTION

In this section we are concerned with the directional derivative of the optimal value function for $P(\varepsilon)$. We first derive upper and lower bounds on the directional derivative limit quotient of $\tilde{f}^*(\varepsilon)$ for $\tilde{P}(\varepsilon)$ and then obtain the corresponding bounds for $P(\varepsilon)$. These results extend the work of Gauvin and Tolle [10], who obtained the analogous results for the case in which the perturbation is restricted to the right-hand side of the constraints.

As above, we will, without loss of generality, focus attention on the parameter value $\varepsilon = 0$. For $z \in E^k$, the directional derivative of $\tilde{f}^*(\varepsilon)$ at $\varepsilon = 0$ in the direction z is defined to be:

$$D_z\tilde{f}^*(0) = \lim_{\beta\to 0^+} \frac{\tilde{f}^*(\beta z) - \tilde{f}^*(0)}{\beta} \tag{11}$$

provided that the limit exists.

THEOREM 7 If, for $\tilde{P}(\varepsilon)$, MFCQ holds for some $\bar{x}_I \in \tilde{S}(0)$, then, for any direction $z \in E^k$,

$$\limsup_{\beta\to 0^+} \frac{\tilde{f}^*(\beta z) - \tilde{f}^*(0)}{\beta} \leqq \max_{\mu\in\tilde{K}(\bar{x}_I,0)} \nabla_\varepsilon \tilde{L}(\bar{x}_I,\mu,0)z \tag{12}$$

PROOF. Let β satisfy the conditions of Theorem 2, let $\delta > 0$ and $\tilde{y}_I$ be the vector given by the constraint qualification, and let $\bar{y}$ satisfy equations (8) and (9). Then, for any $z \in E^k$, $\bar{x}_I + \beta(\bar{y} + \delta\tilde{y}_I) \in \tilde{R}(\beta z)$ for β near 0, so that

$$\limsup_{\beta\to 0^+} \frac{\tilde{f}^*(\beta z) - \tilde{f}^*(0)}{\beta} \leqq \limsup_{\beta\to 0^+} \frac{\tilde{f}(\bar{x}_I + \beta(\bar{y} + \delta\tilde{y}_I),\beta z) - \tilde{f}(\bar{x}_I,0)}{\beta}$$

$$= \frac{d\tilde{f}}{d\beta}(\bar{x}_I,0) = \nabla\tilde{f}(\bar{x}_I,0)(\bar{y} + \delta\tilde{y}_I) + \nabla_\varepsilon\tilde{f}(\bar{x}_I,0)z$$

Since this inequality is satisfied for arbitrary $\delta > 0$ we can take the limit as $\delta \to 0$ and obtain:

$$\limsup_{\beta\to0^+} \frac{\tilde{f}^*(\beta z) - \tilde{f}^*(0)}{\beta} \leqq \nabla\tilde{f}(\bar{x}_I,0)\bar{y} + \nabla_\varepsilon\tilde{f}(\bar{x}_I,0)z$$

The conclusion now follows by applying (9):

$$\limsup_{\beta\to0^+} \frac{\tilde{f}^*(\beta z) - \tilde{f}^*(0)}{\beta} \leqq \max_{\mu\varepsilon\tilde{K}(\bar{x}_I,0)} [-\mu'\nabla_\varepsilon\tilde{g}(\bar{x}_I,0) + \nabla_\varepsilon\tilde{f}(\bar{x}_I,0)]z$$

$$= \max_{\mu\varepsilon\tilde{K}(\bar{x}_I,0)} \nabla_\varepsilon\tilde{L}(\bar{x}_I,\mu,0)z$$

COROLLARY 1 Under the hypotheses of the previous theorem, if MFCQ holds at each point $x_I \varepsilon \tilde{S}(0)$, then

$$\limsup_{\beta\to0^+} \frac{\tilde{f}^*(\beta z) - \tilde{f}^*(0)}{\beta} \leqq \inf_{x_I\varepsilon\tilde{S}(0)} \max_{\mu\varepsilon\tilde{K}(x_I,0)} \nabla_\varepsilon\tilde{L}(x_I,\mu,0)z \qquad (13)$$

PROOF. The result follows directly by applying the previous theorem at each point of $\tilde{S}(0)$.

To obtain a lower bound on the directional derivative limit quotient, we use MFCQ and the following result, which follows easily from the results obtained in the last section.

LEMMA 4 If $\bar{f}^*(\varepsilon)$ is continuous at $\varepsilon = 0$, then $\bar{S}(\varepsilon)$ is closed at $\varepsilon = 0$.

PROOF. Suppose $\varepsilon_n \to 0$ as $n \to \infty$ and suppose $x_I^n \varepsilon \bar{S}(\varepsilon_n)$ is such that $x_I^n \to x_I^*$. By Lemma 2, $\bar{R}(\varepsilon)$ is closed at $\varepsilon = 0$, so $x_I^* \varepsilon \bar{R}(0)$. Since $\bar{f}^*(\varepsilon)$ is continuous at $\varepsilon = 0$, it follows that $\lim_{n\to\infty} \bar{f}^*(\varepsilon_n) = \lim_{n\to\infty} \tilde{f}(x_I^n,\varepsilon_n) = \tilde{f}(x_I^*,0) = \bar{f}^*(0)$; hence $x_I^* \varepsilon \bar{S}(0)$.

The next lemma is an immediate consequence of this result and Theorem 5.

LEMMA 5 Suppose $\varepsilon_n \to 0$ and the assumptions of Theorem 5 hold. Then, for n large, there exists $x_I^n \varepsilon \bar{S}(\varepsilon_n)$ and all limit points of $\{x_I^n\}$ are in $\bar{S}(0)$.

PROOF. The fact that $\bar{S}(\varepsilon) \neq 0$ for ε near 0 follows from the fact that $\bar{R}(\varepsilon_n) \neq \phi$ (Theorem 3) and compact for n large (since $\bar{R}(\varepsilon)$ is uniformly compact for ε near 0) and since $\tilde{f}(x_I,\varepsilon_n)$ is continuous in N*. From Theorem 5, we know that $\bar{f}^*(\varepsilon)$ is continuous at $\varepsilon = 0$. The conclusion then follows from the previous lemma.

DEFINITION 3 For any given vector $z \in E^k$, an infimal sequence x_I^n of the directional derivative limit quotient of $\bar{f}^*(\varepsilon)$ is defined as $\{x_I^n\}$ such that $x_I^n \in \bar{S}(\beta_n z)$ and

$$\liminf_{\beta \to 0^+} \frac{\bar{f}^*(\beta z) - \bar{f}^*(0)}{\beta} = \lim_{n\to\infty} \frac{\tilde{f}(x_I^n, \beta_n z) - \tilde{f}(x_I^*, 0)}{\beta_n}$$

THEOREM 8 Suppose z is any given vector in E^k. Suppose the assumptions of Thoerem 5 are satisfied and suppose that $\bar{x}_I$ is a limit point of an infimal sequence $\{x_I^n\}$ as defined in Definition 3, relative to the given vector z. Then,

$$\liminf_{\beta \to 0^+} \frac{\bar{f}^*(\beta z) - \bar{f}^*(0)}{\beta} \geqq \min_{\mu \in \bar{K}(\bar{x}_I, 0)} \nabla_\varepsilon \tilde{L}(\bar{x}_I, \mu, 0) z \tag{14}$$

PROOF. Let $\beta_n \to 0^+$. We already know from Lemma 5 that there exists $x_I^n \in \bar{S}(\beta_n z)$ for n large, and all limit points of $\{x_I^n\}$ are in $\bar{S}(0)$. Also, by definition, an infimal sequence as defined above always exists and there must exist at least one limit point in $\bar{S}(0)$ of this sequence. Relabeling if necessary, our assumptions allow us to conclude a bit more, i.e., that $x_I^n \to \bar{x}_I \in \bar{S}(0) \cap N^*$, where $\{x_I^n\}$ is an infimal sequence relative to the given vector z.

Since MFCQ holds at $\bar{x}_I$, Theorem 4 assures that $x_I^n + \beta_n(\bar{y} + \delta\tilde{y}_I) \in \tilde{R}(0)$ for n sufficiently large. It follows that

$$\begin{aligned}
\liminf_{\beta \to 0^+} \frac{\bar{f}^*(\beta z) - \bar{f}^*(0)}{\beta} &= \lim_{n\to\infty} \frac{\tilde{f}(x_I^n, \beta_n z) - \tilde{f}(\bar{x}_I, 0)}{\beta_n} \\
&\geqq \lim_{n\to\infty} \frac{\tilde{f}(x_I^n, \beta_n z) - \tilde{f}(x_I^n + \beta_n(\bar{y} + \delta\tilde{y}_I), 0)}{\beta_n} \\
&= \lim_{n\to\infty} [-\nabla\tilde{f}(\alpha_n)(\bar{y} + \delta\tilde{y}_I) + \nabla_\varepsilon \tilde{f}(\alpha_n) z]
\end{aligned}$$

by the mean value theorem, where α_n is the usual convex combination (in $E^{n-p} \times E^k$) of the two arguments in the preceding quotient. Thus,

$$\begin{aligned}
\liminf_{\beta \to 0^+} \frac{\bar{f}^*(\beta z) - \bar{f}^*(0)}{\beta} &\geqq \lim_{n\to\infty} \nabla_\varepsilon \tilde{f}(\alpha_n) z - \nabla\tilde{f}(\alpha_n)(\bar{y} + \delta\tilde{y}_I) \\
&= \nabla_\varepsilon \tilde{f}(\bar{x}_I, 0) z - \nabla\tilde{f}(\bar{x}_I, 0)(y + \delta\tilde{y}_I)
\end{aligned}$$

Using Theorem 1 and noting that δ was chosen as any positive value, we conclude that

$$\liminf_{\beta\to 0^+} \frac{\bar{f}^*(\beta z) - \bar{f}^*(0)}{\beta} \geq \nabla_\varepsilon \tilde{f}(\bar{x}_I,0)z - \max_{\mu\varepsilon\bar{K}(\bar{x}_I,0)} [\mu'\nabla_\varepsilon \tilde{g}(\bar{x}_I,0)z]$$

$$= \min_{\mu\varepsilon\bar{K}(\bar{x}_I,0)} \nabla_\varepsilon \tilde{L}(\bar{x}_I,\mu,0)z$$

COROLLARY 2 Under the hypotheses of the previous theorem,

$$\liminf_{\beta\to 0^+} \frac{\bar{f}^*(\beta z) - \bar{f}^*(0)}{\beta} \geq \inf_{x_I\varepsilon\bar{S}(0)} \min_{\mu\varepsilon\bar{K}(x_I,0)} \nabla_\varepsilon \tilde{L}(x_I,\mu,0)z \qquad (15)$$

By the reduction of variables that was applied earlier, in a neighborhood of $(x_I^*,0)$, with $x = (x_D(x_I,\varepsilon),x_I)$,

$$\begin{aligned} L(x,\mu,\omega,\varepsilon) &= f(x,\varepsilon) - \mu'g(x,\varepsilon) + \omega'h(x,\varepsilon) \\ &= f(x_D(x_I,\varepsilon),x_I,\varepsilon) - \mu'g(x_D(x_I,\varepsilon),x_I,\varepsilon) + \omega'h(x_D(x_I,\varepsilon),x_I,\varepsilon) \\ &= \tilde{f}(x_I,\varepsilon) - \mu'\tilde{g}(x_I,\varepsilon) = \tilde{L}(x_I,\mu,\varepsilon) \end{aligned}$$

with $f(x,\varepsilon) \equiv \tilde{f}(x_I,\varepsilon)$, $g(x,\varepsilon) \equiv \tilde{g}(x_I,\varepsilon)$, and $h(x,\varepsilon) \equiv \tilde{h}(x_I,\varepsilon) \equiv 0$. Thus $L(x,\mu,\omega,\varepsilon) \equiv \tilde{L}(x_I,\mu,\varepsilon)$ in a neighborhood of $(x_D(x_I^*,0),x_I^*,0) = (x^*,0)$ and, with ω determined by $\omega' = -(\nabla_{x_D} f - \mu'\nabla_{x_D} g)[\nabla_{x_D} h]^{-1}$, it follows easily that $\nabla_\varepsilon \tilde{L} = \nabla_\varepsilon L$ and the linear program appearing in the proof of Theorem 1 and involved in the preceding bounds can readily be formulated analogously as a locally equivalent problem in terms of $L(x,\mu,\omega,\varepsilon)$.

We now utilize the above results obtained for $\tilde{P}(\varepsilon)$ and $\bar{P}(\varepsilon)$ to obtain bounds for the optimal value directional derivative quotient of $P(\varepsilon)$.

THEOREM 9 If MFCQ holds at some $x^* \varepsilon S(0)$, then for any direction $z \varepsilon E^k$,

$$\limsup_{\beta\to 0^+} \frac{f^*(\beta z) - f^*(0)}{\beta} \leq \max_{(\mu,\omega)\varepsilon K(x^*,0)} \nabla_\varepsilon L(x^*,\mu,\omega,0)z$$

PROOF. Apply the variable reduction transformation at $x^* = (x_D^*,x_I^*)$ as in the previous construction to obtain a problem of the form $\tilde{P}(\varepsilon)$, defined for (x_I,ε) in a neighborhood N^* of $(x_I^*,0)$. Since $f^*(\varepsilon) \leq \tilde{f}^*(\varepsilon)$ and $f^*(0) = \tilde{f}^*(0)$, we have that

$$\limsup_{\beta\to 0^+} \frac{f^*(\beta z) - f^*(0)}{\beta} \leq \limsup_{\beta\to 0^+} \frac{\tilde{f}^*(\beta z) - \tilde{f}^*(0)}{\beta}$$

and the conclusion is an immediate consequence of Theorem 7, having expressed in (12) in terms of the original variables by way of the variable reduction transformation.

THEOREM 10 If $R(0) \neq \phi$, $R(\varepsilon)$ is uniformly compact near $\varepsilon = 0$, and MFCQ holds for each $x \in S(0)$, then for any direction $z \in E^k$,

$$\liminf_{\beta \to 0^+} \frac{f^*(\beta z) - f^*(0)}{\beta} \geqq \min_{(\mu,\omega) \in K(x^*,0)} \nabla_\varepsilon L(x^*,\mu,\omega,0) z$$

holds for some $x^* \in S(0)$.

PROOF. The fact that $S(\varepsilon) \neq \phi$ for large n follows from an application of the variable reduction transformation at any $\bar{x} \in S(0)$, yielding a problem of the form $\bar{P}(\varepsilon)$ where $(\bar{x}_I, 0) \in M^0$, and then applying Lemma 5.

Given any $z \in E^n$, consider $x^n \in S(\beta_n z)$ such that

$$\liminf_{\beta \to 0^+} \frac{f^*(\beta z) - f^*(0)}{\beta} = \lim_{n \to \infty} \frac{f(x^n, \beta_n z) - f(x^*, 0)}{\beta_n}$$

Since $R(\varepsilon)$ is uniformly compact, there exists a subsequence, which we again denote by $\{x^n\}$, and a vector x^* such that $x_n \to x^*$. By Lemma 3, $R(\varepsilon)$ is closed and, by Theorem 6, $f^*(\varepsilon)$ is continuous at $\varepsilon = 0$. It follows (as in the proof of Lemma 4) that $S(\varepsilon)$ is closed at 0, so $x^* \in S(0)$.

We now apply the variable reduction transformation at $x^* = (x_D^*, x_I^*)$, following the usual construction, and obtain a problem of the form $\tilde{P}(\varepsilon)$, defined for (x_I, ε) in a neighborhood N^* of $(x_I^*, 0)$. We also define the reduced problem $\bar{P}(\varepsilon)$, i.e., problem $\tilde{P}(\varepsilon)$ with M, a closed subset of N^* whose interior contains $(x_I^*, 0)$, replacing N^*.

Let $x_n = (x_D^n, x_I^n)$. Then $(x_I^n, \beta_n z) \to (x_I^*, 0)$ as $n \to \infty$. For n large enough, $(x_I^n, \beta_n z) \in M$, implying also that $x_D^n = x_D(x_I^n, \beta_n z)$, where $x_D(x_I, \varepsilon)$ is the unique continuous vector function defined on N^* that results from the variable reduction at x^*. We conclude that $f^*(\beta_n z) \equiv f(x_n, \beta_n z) = f[x_D(x_I^n, \beta_n z), x_I^n, \beta_n z] \equiv \tilde{f}(x_I^n, \beta_n z) \geqq \bar{f}^*(\beta_n z)$, where $\bar{f}^*(\beta_n z)$ is the optimal value of $\bar{P}(\beta_n z)$, the last inequality following from the fact that $0 \leqq g(x_n, \beta_n z) = g[x_D(x_I^n, \beta_n z), x_I^n, \beta_n z] \equiv \tilde{g}(x_I^n, \beta_n z)$, so that x_I^n is a feasible point of $\bar{P}(\beta_n z)$. Since $f^*(\beta_n z) \leqq \bar{f}^*(\beta_n z)$, it follows that $f^*(\beta_n z) = \bar{f}^*(\beta_n z)$ in N^*.

It is easily verified that the assumptions of Theorem 8 are satisfied and we also have

$$\liminf_{\beta \to 0^+} \frac{f^*(\beta z) - f^*(0)}{\beta} = \liminf_{\beta \to 0^+} \frac{\bar{f}^*(\beta z) - \bar{f}^*(0)}{\beta}$$

from which the conclusion follows, expressing the right-hand side of (14) in terms of the original variables via the variable reduction transformation.

Clearly Corollaries 1 and 2 immediately extend to $f^*(\varepsilon)$ as well, using these results. Thus, all of the results obtained above for $\tilde{P}(\varepsilon)$ and $\bar{P}(\varepsilon)$ can be immediately generalized to $P(\varepsilon)$. For completeness, we state these results as the next theorem.

THEOREM 11 If, for $P(\varepsilon)$, $R(0)$ is nonempty and MFCQ holds at each $x \in S(0)$, then for any unit vector $z \in E^k$,

$$\limsup_{\beta\to 0^+} \frac{f^*(\beta z) - f^*(0)}{\beta} \leqq \inf_{x\in S(0)} \max_{(\mu,\omega)\in K(x,0)} \nabla_\varepsilon L(x,\mu,\omega,0)z \tag{16}$$

and if $R(\varepsilon)$ is uniformly compact for ε near $\varepsilon = 0$, then

$$\liminf_{\beta\to 0^+} \frac{f^*(\beta z) - f^*(0)}{\beta} \geqq \inf_{x\in S(0)} \min_{(\mu,\omega)\in K(x,0)} \nabla_\varepsilon L(x,\mu,\omega,0)z \tag{17}$$

Moreover, we are able to obtain the existence of the directional derivative of f^* at $\varepsilon = 0$ by assuming, as Gauvin and Tolle [10] did for right-hand-side programs, the linear independence of the binding constraint gradients at each point $x^* \in S(0)$.

COROLLARY 3 Assume $R(0)$ is nonempty and $R(\varepsilon)$ is uniformly compace near $\varepsilon = 0$. If the gradients, taken with respect to x, of the constraints binding at x^* are linearly independent for each $x^* \in S(0)$, then for any unit vector $z \in E^k$, $D_z f^*(0)$ exists and is given by

$$D_z f^*(0) = \inf_{x\in S(0)} \nabla_\varepsilon L(x,\mu(x),\omega(x),0)z$$

where $(\mu(x),\omega(x))$ is the unique multiplier vector associated with x.

PROOF. At any point $x^* \in S(0)$, the linear independence of the binding constraint gradients implies the uniqueness of the Kuhn-Tucker multipliers corresponding to x^*. Inequalities (16) and (17) now combine to yield the desired result.

Gauvin and Dubeau [9] showed that the inf in (17) *can* be replaced by min and have examples that show that the inf in (16) *cannot* be replaced by min.

Note that in Corollary 3 if $P(\varepsilon)$ contains no inequality constraints, we could replace inf by min since μ would not appear and $\omega' = -\nabla_{x_D} f \, \nabla_{x_D} h^{-1}$, which is continuous in x, making $\nabla_\varepsilon Lz$ a continuous function of x minimized over S(0), a compact set.

We may also show that two of the observations made by Gauvin and Tolle [10] about $D_z f^*(0)$ for right-hand-side programs apply to P(ε) as well. First if $D_z f^*(0) = -D_{-z} f^*(0)$, then

$$\inf_{x\varepsilon S(0)} \max_{(\mu,\omega)\varepsilon K(x,0)} \nabla_\varepsilon L(x,\mu,\omega,0)z = \sup_{x\varepsilon S(0)} \min_{(\mu,\omega)\varepsilon K(x,0)} \nabla_\varepsilon L(x,\mu,\omega,0)z \tag{18}$$

Thus, if, for all unit vectors $z \varepsilon E^k$, $D_z f^*(0) - -D_{-z} f^*(0)$ and

$$D_z f^*(0) = \inf_{x\varepsilon S(0)} \max_{(\mu,\omega)\varepsilon K(x,0)} \nabla_\varepsilon L(x,\mu,\omega,0)z \tag{19}$$

then (18) provides a necessary condition for the existence of $\nabla_\varepsilon f^*(0)$. In addition, if (18) holds for every unit vector $z \varepsilon E^k$ and if $x^* \varepsilon S(0)$ is the unique solution of P(0), then its associated Kuhn-Tucker multiplier vector is unique.

We next apply the results derived above to a particular class of programs. We show in the next theorem that if P(ε) is a convex program in x for ε near ε = 0, i.e., if f(x,ε) and $-g_i(x,\varepsilon)$, i = 1,...,m, are convex and if $h_j(x,\varepsilon)$, j = 1,...,p, are affine in x, then $D_z f^*(0)$ exists and is given by (19). To prove this result, we will restrict our attention to convex programs of the form $\tilde{P}(\varepsilon)$, with (x_I,ε) no longer constrained to be in a specified neighborhood N*. We are able to do this since the functions $h_j(x,\varepsilon)$ are assumed to be affine in x and f(x) and the $-g_i(x,\varepsilon)$ are taken to be convex in x, from which it easily follows that the variable reduction transformation applies globally, and further, $\tilde{f}(x_I,\varepsilon)$ and the $-\tilde{g}_i(x_I,\varepsilon)$ are convex in x_I for i = 1,...,m.

THEOREM 12 In P(ε), let f(x,ε) and $-g_i(x,\varepsilon)$, i = 1,...,m, be convex and let $h_j(x,\varepsilon)$, j = 1,...,p, be affine in x. If R(0) is nonempty, R(ε) is uniformly compact near ε = 0, and MFCQ holds for each $x^* \varepsilon S(0)$, then, for any unit vector $z \varepsilon E^k$,

$$D_z f^*(0) = \inf_{x\varepsilon S(0)} \max_{(\mu,\omega)\varepsilon K(x,0)} \nabla_\varepsilon L(x,\mu,\omega,0)z \tag{20}$$

PROOF. Without loss of generality, as indicated, we will prove this result for $\tilde{P}(\varepsilon)$, with the set N* not present. For convenience, the notation is somewhat simplified by dropping the subscript I in the argument. Note that the assumptions imply that $\tilde{R}(\varepsilon) \equiv \{x \mid \tilde{g}_i(x,\varepsilon) \geqq 0,\ i = 1,\ldots,m\}$ is a convex uniformly compact set near $\varepsilon = 0$.

Let $x^* \in \tilde{S}(0)$ and $x_n \in \tilde{S}(\beta_n z)$ with $\beta_n \to 0^+$ in such a way that

$$\liminf_{\beta\to 0^+} \frac{\tilde{f}^*(\beta z) - \tilde{f}^*(0)}{\beta} = \lim_{n\to\infty} \frac{\tilde{f}(x_n,\beta_n z) - \tilde{f}(x^*,0)}{\beta_n}$$

and $x_n \to x^*$ as in the proof of Theorem 10. For all $\mu^* \in \tilde{K}(x^*,0)$,

$$\tilde{L}(x_n,\mu^*,\beta_n z) = \tilde{f}(x_n,\beta_n z) - \mu^{*\prime}\tilde{g}(x_n,\beta_n z) \leqq \tilde{f}(x_n,\beta_n z)$$

where the inequality follows from the nonnegativity of both μ^* and $\tilde{g}(x_n,\beta_n z)$. Thus, since $\tilde{L}(x^*,\mu^*,0) = f(x^*,0)$,

$$\lim_{n\to\infty} \frac{\tilde{f}(x_n,\beta_n z) - \tilde{f}(x^*,0)}{\beta_n} \geqq \lim_{n\to\infty} \frac{\tilde{L}(x_n,\mu^*,\beta_n z) - \tilde{L}(x^*,\mu^*,0)}{\beta_n}$$

Now, as a result of the Kuhn-Tucker conditions and the convexity assumptions, x^* is a global minimizer of $\tilde{L}(x,\mu^*,0)$, so

$$\lim_{n\to\infty} \frac{\tilde{L}(x_n,\mu^*,\beta_n z) - \tilde{L}(x^*,\mu^*,0)}{\beta_n} \geqq \lim_{n\to\infty} \frac{\tilde{L}(x_n,\mu^*,\beta_n z) - \tilde{L}(x_n,\mu^*,0)}{\beta_n}$$

$$= \lim_{n\to\infty} \frac{\tilde{L}(x_n,\mu^*,0) + \beta_n \nabla_\varepsilon \tilde{L}(x_n,\mu^*,t\beta_n z)z - \tilde{L}(x_n,\mu^*,0)}{\beta_n}$$

by the mean value theorem, where $t \in (0,1)$. Thus,

$$\lim_{n\to\infty} \frac{\tilde{f}(x_n,\beta_n z) - \tilde{f}(x^*,0)}{\beta_n} \geqq \lim_{n\to\infty} \nabla_\varepsilon \tilde{L}(x_n,\mu^*,t\beta_n z)z$$

and, passing to the limit on the right, we are able to conclude that

$$\lim_{n\to\infty} \frac{\tilde{f}(x_n,\beta_n z) - \tilde{f}(x^*,0)}{\beta_n} \geqq \nabla_\varepsilon \tilde{L}(x^*,\mu^*,0)z \tag{21}$$

Thus, for some $x^* \in \tilde{S}(0)$, since (21) holds for each $\mu \in \tilde{K}(x^*,0)$, and recalling from [10] that $\tilde{K}(x^*,0)$ is compact,

$$\liminf_{\beta\to 0^+} \frac{\tilde{f}^*(\beta z) - \tilde{f}^*(0)}{\beta} \geqq \max_{\mu\in\tilde{K}(x^*,0)} \nabla_\varepsilon \tilde{L}(x^*,\mu,0)z$$

from which we see that

$$\liminf_{\beta\to 0^+} \frac{\tilde{f}^*(\beta z) - \tilde{f}^*(0)}{\beta} \geqq \inf_{x\in\tilde{S}(0)} \max_{\mu\in\tilde{K}(x,0)} \nabla_\varepsilon \tilde{L}(x,\mu,0)z$$

Combining this result with that obtained in Corollary 1 we conclude that

$$D_z\tilde{f}^*(0) = \inf_{x\in\tilde{S}(0)} \max_{\mu\in\tilde{K}(x,0)} \nabla_\varepsilon \tilde{L}(x,\mu,0)z$$

For convex $P(\varepsilon)$, (20) now follows by an inversion of the reduction of variables process applied to yield $\tilde{P}(\varepsilon)$.

V. EXAMPLE

We use the example stated below to demonstrate some of the theoretical results obtained in the previous sections. For the given problem we show that the conditions of MFCQ hold at every point in $S(\varepsilon)$ and we give the form of the vector satisfying the constraint qualification. We obtain the form of the vector satisfying (1) and (2) for $\tilde{P}(\varepsilon)$, and show that the bounds stated in (16) and (17) are attained.

Consider the program

$$\begin{aligned} &\min \ \varepsilon x_1 \\ &\text{s.t.} \ \ g(x,\varepsilon) = -(x_1 - \varepsilon)^2 - (x_2 - 2)^2 + 4 \geqq 0 \\ &\qquad h(x,\varepsilon) = -x_1 + x_2 + \varepsilon = 0 \end{aligned} \qquad P(\varepsilon)$$

The solution of this program is easily determined to be $x_1^* = x_2^* + \varepsilon$ with

$$x_2^* = \begin{cases} 0 & \varepsilon > 0 \\ 2 & \varepsilon < 0 \end{cases} \quad \text{and if } \varepsilon = 0,\ x_2^* \text{ can be any value in the interval } [0,2] \qquad (22)$$

Applying the reduction of variables technique outlined earlier, with $x_D = x_1$ and $x_I = x_2$, $P(\varepsilon)$ is transformed into the equivalent program

$$\begin{aligned} &\min \ \varepsilon(x_2 + \varepsilon) \\ &\text{s.t.} \ \ \tilde{g}(x_2,\varepsilon) = -x_2^2 - (x_2 - 2)^2 + 4 \geqq 0 \end{aligned} \qquad \tilde{P}(\varepsilon)$$

whose solution is given by (22).

For both $P(\varepsilon)$ and $\tilde{P}(\varepsilon)$, the optimal value function can be written as

$$f^*(\varepsilon) = \begin{cases} \varepsilon^2 & \varepsilon \geqq 0 \\ \varepsilon^2 + 2\varepsilon & \varepsilon < 0 \end{cases} \qquad (23)$$

We see that f^* is continuous for all values of ε, but it is not differentiable at $\varepsilon = 0$. It does, however, have directional derivatives at $\varepsilon = 0$ which are given by

$$D_z f^*(0) = \begin{cases} 0 & z = 1 \\ -2 & z = -1 \end{cases} \qquad (24)$$

To illustrate Lemma 1, we first determine the general form of the vector, $\tilde{y}$, associated with points $x \in S(\varepsilon)$ at which MFCQ is satisfied. The constraint gradients of $P(\varepsilon)$ are

$$\nabla g(x,\varepsilon) = [-2(x_1 - \varepsilon), -2(x_2 - 2)] \qquad \text{and}$$

$$\nabla h(x,\varepsilon) = [-1,1]$$

Applying (1) and (2) at a point $x^* = (x_1^*, x_2^*) \in S(\varepsilon)$, with $\tilde{y} = (y_1, y_2)$, we require that $\nabla g(x^*,\varepsilon)\tilde{y} = -2(x_1^* - \varepsilon)y_1 - 2(x_2^* - 2)y_2 > 0$ if $g(x^*,0) = 0$, and

$$\nabla h(x^*,\varepsilon)\tilde{y} = -y_1 + y_2 = 0$$

Thus, for any value of ε, since $g(x,\varepsilon)$ is binding only if $x_2^* = 0,2$, $\tilde{y}$ can have the form

$$\tilde{y} = \begin{cases} (a,a) & x_2^* = 0 \\ (b,b) & 0 < x_2^* < 2 \\ (c,c) & x_2^* = 2 \end{cases} \tag{25}$$

for any real numbers a,b,c with $a > 0$, $b \neq 0$, and $c < 0$. We can also conclude that MFCQ holds at every solution of $P(\varepsilon)$.

In a similar fashion, we see that, for $\tilde{P}(\varepsilon)$,

$$\nabla\tilde{g}(x_2,\varepsilon) = -2x_2 - 2(x_2 - 2)$$

so applying (1) we find that the vector $\tilde{y}_I$ in the reduced program takes the same form as the second component of $\tilde{y}$ in (25).

Now

$$\nabla L(x,\mu,\omega,\varepsilon) = [\varepsilon + 2\mu(x_1 - \varepsilon) - \omega,\ 2\mu(x_2 - 2) + \omega]$$

so that at a solution $x^* \in S(0)$ we must have $2\mu x_1^* - \omega = 0$ for $(\mu,\omega) \in K(x^*,0)$. Then,

$$\nabla_\varepsilon L(x^*,\mu,\omega,0) = x_1^*$$

and, with $S(0) = \{x \in E^2 : x_1 = x_2,\ x_2 \in [0,2]\}$,

$$\min_{x \in S(0)} \ \max_{(\mu,\omega) \in K(x,0)} \nabla_\varepsilon L(x,\mu,\omega,0)z = \begin{cases} 0 & z = 1 \\ -2 & z = -1 \end{cases} \tag{26}$$

Comparing (24) with (26) we see that (16) holds with equality.

Now, considering inequality (17), we first note that for any neighborhood $N(0)$ of $0 \in E^1$, the closure of the set

$$\{x \in E^2 : x = (x_2 + \varepsilon, x_2), x_2 \in [0,2], \varepsilon \text{ in } N(0)\}$$

is compact so $R(\varepsilon)$ is uniformly compact for ε near $\varepsilon = 0$. We calculate

$$\min_{x\varepsilon S(0)} \min_{(\mu,\omega)\varepsilon K(x,0)} \nabla_\varepsilon L(x,\mu,\omega,0)z = \begin{cases} 0 & z = 1 \\ -2 & z = -1 \end{cases}$$

and find that (17) also holds with equality.

The above results could have been anticipated from (20), since the conditions of Theorem 12 hold for this example.

An example is given in [10] that illustrates that (16) and (17) need not hold with equality.

VI. RELATED RESULTS

Inspection of the derivation of (16) and (17) reveals that the bounding term in these expressions, namely, $\nabla_\varepsilon L(x,\mu,\omega,0)z$, can be viewed as the sum of two distinct expressions, one resulting from the variation of the objective function of $P(\varepsilon)$ with respect to the parameter, the other deriving from the dependence of the region of feasibility on the parameter. The first of these terms is $\nabla_\varepsilon f(x,0)z$ and is easily seen to result directly from the manipulation of the limit quotients in the proofs of Theorems 7 and 8. The second component, $[-\mu'\nabla_\varepsilon g(x,0) + \omega'\nabla_\varepsilon h(x,\varepsilon)]z$, results from the assumption that MFCQ holds at points of $S(0)$. The conditions of MFCQ are invoked to enable us to conclude (9), as well as the existence of points feasible to $P(\varepsilon)$ in a neighborhood of $\varepsilon = 0$. Having made these observations, we are now able to discuss the relationships between the bounds provided here and results previously obtained by others. As we shall see, in particular instances in which the directional derivative of f^* is shown to exist, it is expressed as either a function of $\nabla_\varepsilon f$ or a function of $\nabla_\varepsilon g$ and $\nabla_\varepsilon h$, or a combination of all of these terms, depending, as one would suspect, on where in $P(\varepsilon)$ the parameter appears.

For convex programs, the existence of $D_z f^*(0)$ assured by Theorem 12 and its expression as (20), corresponds under slightly different assumptions, with results achieved by Gol'stein [12] and Hogan [14]. Theorem 12 is an extension to the general perturbed mathematical program of a result given by Gauvin and Tolle [10] for right-hand-side programs.

It is also relevant to note that Geraud Fontanie [8] extended the Gauvin-Tolle bounds [10] to a generally perturbed Lipschitz program, using the reduction technique described here.

As mentioned in the introduction, Rockafellar [20] has recently provided sharper results under considerably weaker assumptions than those given here and, for that matter, for related results in the bulk of the recent literature on this subject. He thoroughly explores the relationship between generalized Lagrange multipliers and directional derivatives, subgradients and subderivatives of the optimal value function of the general parametric programming problem, using the concepts of subdifferential analysis. He develops a unified and more general theoretical framework for optimal value differential stability results.

Rockafellar [20, p. 4] asserts that the results of Gauvin and Tolle [10] can be immediately extended to the general parametric problem $P(\varepsilon)$ by applying their results to a reformulation of this problem as the equivalent right-hand-side perturbation problem

$$\begin{array}{lll} \min_{(x,\varepsilon)} & f(x,\varepsilon) & \\ \text{s.t.} & g_i(x,\varepsilon) \geqq 0 & (i = 1,\ldots,m) \\ & h_j(x,\varepsilon) = 0 & (j = 1,\ldots,p) \\ & \varepsilon = \alpha & \end{array} \qquad \bar{P}(\alpha)$$

where the parameter is now α. This idea was also suggested to the author by Janin [16], who pointed out that the MFCQ constraint qualification is inherited from $P(\varepsilon)$ by $\bar{P}(\alpha)$. This latter fact is easily proved and the main property that need be observed to conclude the validity of the assertion: the results on the continuity and Dini derivative bounds for the optimal value function $f^*(\varepsilon)$ of problem $P(\varepsilon)$ are implied by the results of Gauvin and Tolle [10] via the problem $\bar{P}(\alpha)$. It is remarkable that this was not noticed earlier.

Thus, the results we have given here are, like those of Gauvin and Dubeau [9], actually implied by those of Gauvin and Tolle [10]. Nonetheless, we submit again that the virtue of the results presented herein is their directness, theoretical simplicity, and amenability to clearer interpretation and hence computational verification and numerical exploitation than those provided heretofore. The reduction approach simplifies the problem statement and makes possible a direct constructive treatment by elementary arguments.

We are grateful to a referee for pointing out that Gollan [11] uses $\bar{P}(\alpha)$ to prove a general result for $P(\varepsilon)$.

REFERENCES

1. A. Auslender. Differential stability in nonconvex and nondifferentiable programming, *Mathematical Programming Study 10*, pp. 29-41, North Holland, 1979.

2. C. Berge. *Topological Spaces* (translated by E. M. Patterson). Macmillan, New York, 1963.

3. J. M. Danskin. *The Theory of Max-Min*. Springer-Verlag, New York, 1967.

4. V. E. Dem'janov and A. B. Pevnyi. First and second marginal values of mathematical programming problems, *Soviet Math. Dokl.*, Vol. 13, pp. 1502-1506 (1972).

5. A. V. Fiacco. Continuity of the optimal value function under the Mangasarian-Fromovitz constraint qualification, Technical Paper Serial T-432, Institute for Management Science and Engineering, The George Washington University (1980).

6. A. V. Fiacco and W. P. Hutzler. Extensions of optimal value differential stability results to general mathematical programs, Technical Paper Serial T-393, Institute for Management Science and Engineering, The George Washington University (1979). (A version of this paper, entitled Optimal value differential stability results for general inequality constrained differentiable mathematical programs, was published in *Mathematical Programming with Data Perturbations*, Vol. I (A. V. Fiacco, ed.), *Lecture Notes in Pure and Applied Mathematics*, Vol. 73, pp. 29-43, Marcel Dekker, 1982.)

7. A. V. Fiacco and W. P. Hutzler. Basic results in the development of sensitivity and stability analysis in nonlinear programming, Technical Paper Serial T-407, Institute for Management Science and Engineering, The George Washington University (1979).

8. G. Fontanie. Locally Lipschitz functions and nondifferentiable programming. Master's thesis; Technical Report 80-3, Curriculum in Operations Research and Systems Analysis, University of North Carolina at Chapel Hill (1980).

9. J. Gauvin and F. Dubeau. Differential properties of the marginal function in mathematical programming, *Mathematical Programming Study* (M. Guignard, ed.), to appear.

10. J. Gauvin and J. W. Tolle. Differential stability in nonlinear programming, *SIAM Journal of Control and Optimization*, Vol. 15, pp. 294-311 (1977).

11. B. Gollan. A general perturbation theory for abstract optimization problems, *Journal of Optimization Theory and Applications*, Vol. 35, pp. 417-441 (1981).

12. E. G. Gol'stein. *Theory of Convex Programming*. Translation of *Mathematical Monographs*, Vol. 36, American Mathematical Society, Providence, Rhode Island, 1972.

13. M. R. Hestenes. *Optimization Theory: The Finite Dimensional Case*. Wiley, New York, 1975.

14. W. Hogan. Directional derivatives for extremal value functions with applications to the completely convex case, *Operations Research*, Vol. 21, pp. 188-209 (1973).

15. W. Hogan. Point-to-set maps in mathematical programming, *SIAM Review*, Vol. 15, pp. 591-603 (1973).

16. R. Janin. Personal communication, January 1981.

17. F. Lempio and H. Maurer. Differential stability in infinite dimensional nonlinear programming, *Applied Math and Optimization*, Vol. 6, pp. 139-152 (1980).

18. E. S. Levitin. Differentiability with respect to a parameter of the optimal value in parametric problems of mathematical programming, *Cybernetics*, Vol. 12, pp. 46-60 (1976).

19. O. L. Mangasarian. *Nonlinear Programming*. McGraw-Hill, New York, 1969.

20. R. T. Rockafellar. Lagrange multipliers and subderivatives of optimal value functions in nonlinear programming, *Mathematical Programming Study 17* (R. Wets, ed.), to appear.

CHAPTER 6 ITERATION AND SENSITIVITY FOR A NONLINEAR SPATIAL EQUILIBRIUM PROBLEM*

CAULTON L. IRWIN and CHIN W. YANG† / West Virginia University, Morgantown, West Virginia

ABSTRACT

An iterative solution method and sensitivity analysis are considered for a multi-product, multi-regional, spatial equilibrium problem. Nonlinear supply and demand functions are treated via successive linear approximations. Both the sensitivity analysis and the iteration convergence are based upon a solvability lemma, which guarantees that a solution of the equilibrium conditions may be expressed in terms of the problem data. Convergence of the iterative method is also proved by relating the equilibrium conditions to a nonlinear complementarity problem.

I. INTRODUCTION

In a talk on general observations on NLP methodology at the first Symposium on Mathematical Programming with Data Perturbations, A. V. Fiacco commented that algorithm convergence is closely related to sensitivity and stability analysis. This point is also illustrated by McCormick in [8], as an appli-

---

*Revised April 22, 1981.

†*Current affiliation:* Clarion State College, Clarion, Pennsylvania

cation of second order sufficiency conditions. In this paper, which was motivated by a spatial, supply-demand model of Eastern coal markets, the close connection of algorithm convergence and sensitivity analysis is further illustrated and analyzed.

Our approach to sensitivity and the convergence of iterative solution methods for the nonlinear spatial equilibrium problem in this paper is via local linear approximations. In a previous paper, Irwin and Yang [6], the case of sensitivity and convergence for linear supply and demand functions was discussed.

Section II of this paper is a statement and discussion of the spatial equilibrium conditions. Section III presents a solvability condition that guarantees that an equilibrium solution may be expressed uniquely in terms of the problem parameters and data. An iterative solution method and its convergence are discussed in Section IV, while the sensitivity analysis of an equilibrium solution is considered in Section V.

II. STATEMENT OF EQUILIBRIUM CONDITIONS

We assume that P_S is a function from R^m into R^m with component functions, P_{S_i}, $i = 1,\ldots,m$, which are possibly nonlinear and that P_D is a function from R^n into R^n with possibly nonlinear component functions. The functions P_{S_i} may be thought of as expressing the dependence of the supply price of the ith resource upon the production levels, x_k, $k = 1,\ldots,m$, of all resources in all supply regions. Similarly, P_{D_j} expresses the dependence of the demand price of the jth product upon the consumption levels, y_ℓ, $\ell = 1, \ldots,n$, of all products in all demand regions.

Assume for now that there are no transshipment nodes between the m sources and the n sinks. Let the vector z in R^{mn} have components z_{ij} equal to the quantities shipped from supply region i to demand region j and, for a given shipment pattern z, let $t_{ij}(z)$ denote the unit shipping costs from source i to sink j.

The spatial equilibrium problem as presented, for example, in Takayama and Judge [13], Kennedy [7], and Silberberg [11], may now be stated. Determine supply levels $\{x_i\}_{j=1}^m$, supply prices $\{\gamma_j\}_{j=1}^m$, demand levels $\{y_i\}_{i=1}^n$, demand prices $\{\lambda_j\}_{j=1}^n$, and interregional flows $\{z_{ij}\}_{i=1,j=1}^{m,n}$ so that, in vector notation, the following spatial equilibrium conditions hold:

$$\gamma \leqq P_S(x) \qquad x^T[P_S(x) - \gamma] = 0 \tag{1}$$

$$\lambda \geqq P_D(y) \qquad y^T[\lambda - P_D(y)] = 0 \tag{2}$$

$$G_y^T\lambda \leqq G_x^T\gamma + t(z) \qquad [\gamma^T G_x + t(z)^T - \lambda^T G_y]z = 0 \tag{3}$$

$$G_x z \leqq x \qquad \gamma^T[x - G_x z] = 0 \tag{4}$$

$$G_y z \geqq y \qquad \lambda^T[G_y z - y] = 0 \tag{5}$$

$$x \geqq 0,\ y \geqq 0,\ z \geqq 0,\ \gamma \geqq 0,\ \lambda \geqq 0 \tag{6}$$

The matrix G_x in (4) is defined so that the resource constraints $\sum_{j=1}^{n} \alpha_{ij} z_{ij} \leqq x_i$ hold for i = 1,...,m. Similarly, the matrix G_y in (5) is used to express the conditions that for j = 1,...,n, $\sum_{i=1}^{m} \beta_{ij} z_{ij} \geqq y_j$, i.e., demand levels are met in the n demand regions. We require that $\alpha_{ij} > 0$ and $\beta_{ij} > 0$ for $1 \leqq i \leqq m$ and $1 \leqq j \leqq n$.

The economic interpretations of conditions (1-6) are very interesting and are particularly well described in Chapter 12 of Takayama and Judge [13].

Now, in case the functions P_S, P_D, and t are gradients, i.e., there exist scalar valued functions f, g, and τ so that $\nabla f = P_S$, $\nabla g = P_D$, and $\nabla\tau = t$, the equilibrium conditions (1-6) are the Kuhn-Tucker necessary conditions that must be satisfied by a solution of the nonlinear programming problem

$$\begin{aligned} &\max_{x,y,z} \ \{g(y) - f(x) - \tau(z)\} \\ &\text{s.t.} \quad G_x z \leqq x \\ &\qquad\ G_y z \geqq y \\ &\qquad\ x,y,z \geqq 0 \end{aligned} \tag{7}$$

The gradient condition, which enables one to formulate (7), requires rather strong assumptions on P_S, P_D, and t; this condition is also known as the integrability condition. The maximization form of the spatial equilibrium problem leads to quasi-welfare (i.e., consumers' plus producers' surplus) interpretations of economic equilibria; see Chapter 7 of Takayama and Judge [13].

Our interest, in this paper, is in the solution by iterative methods and a sensitivity analysis of (1-6) whenever there is *not* an equivalent maximization form of the problem. This nonintegrable, i.e., nongradient situation could occur in an econometrically determined demand model where the cross price effects are not necessarily equal. For example, the nonintegrable case holds if P_D is the linear function $P_D(y) = C + Dy$ and D is not a symmetric matrix. Furthermore, a constant elasticity demand model is integrable only if the cross price elasticities are zero.

III. SOLVABILITY CONDITION

The solvability condition guarantees that a solution of (1-6) may be expressed in terms of the problem parameters and data. This is obviously useful in a sensitivity analysis, but the solvability condition also enables one to prove convergence for certain iterative solution methods. Some results that are stated and proved in [6] will be used in the present analysis and are stated in the following lemmas. For these lemmas, assume that $P_S(x) > 0$ for all $x \geqq 0$, i.e., there are no products provided at zero cost. Also assume that transportation costs are nonnegative, i.e., $t(z) \geqq 0$ for all shipment schedules z.

Let x, y, z, γ, λ denote a solution of (1-6) and let

$$\bar{G} = \begin{pmatrix} \bar{G}_x \\ --- \\ \bar{G}_y \end{pmatrix} \text{ denote the columns of } G = \begin{pmatrix} G_x \\ --- \\ G_y \end{pmatrix}$$

which correspond to positive z_{ij}; $\bar{G}$ is an $(m + n) \times k$ matrix of rank r.

LEMMA 1 If x, y, z, γ, λ is a solution of (1-6), then

$$G_x z = x \tag{8}$$

and

$$G_y z = y \tag{9}$$

LEMMA 2 A solution of (1-6) satisfies the equation

$$-P_S(x)^T \bar{G}_x + P_D(y)^T \bar{G}_y = \bar{t}(z)^T \tag{10}$$

The result in Lemma 1 implies that $\begin{pmatrix} x \\ - \\ y \end{pmatrix}$ is in the column space of $\bar{G}$, which means that $\begin{pmatrix} x \\ - \\ y \end{pmatrix}$ is in the orthogonal complement of the null space of $\bar{G}^T$. If we let $\bar{H}$ be an $(m + n - r) \times (m + n)$ matrix whose rows span the null space of $\bar{G}^T$, then

$$\bar{H}\begin{pmatrix} x \\ - \\ y \end{pmatrix} = 0 \tag{11}$$

Equations (10) and (11) constitute a system of $m + n + (k - r)$ (possibly nonlinear) equations in x and y, which is very useful for convergence and sensitivity analysis.

Another important result from [6] applies to the case in which transportation costs are constant, i.e., $t(z) = t$ and P_S and P_D are affine functions, i.e., $P_S(x) = C + Sx$ and $P_D(y) = D + Uy$. Making these substitutions into (10) and (11) we obtain the system

$$\begin{pmatrix} \left(\bar{G}_x^T \mid \bar{G}_y^T\right) \begin{pmatrix} -S & 0 \\ 0 & U \end{pmatrix} \\ \hline \bar{H} \end{pmatrix} \begin{pmatrix} x \\ \hline y \end{pmatrix} = \begin{pmatrix} \bar{t} + \left(\bar{G}_x^T \mid \bar{G}_y^T\right) \begin{pmatrix} C \\ \hline -D \end{pmatrix} \\ \hline 0 \end{pmatrix} \tag{12}$$

This system is of fundamental importance in our sensitivity and iteration by virtue of the following lemmas, which are also proved in [6].

LEMMA N If x, y, z, γ, λ is a solution of (1-6) with $t(z) = t$, $P_S(x) = C + Sx$, and $P_D(y) = D + Uy$, then x and y must satisfy the system of equations (12).

LEMMA S If $\begin{pmatrix} -S & 0 \\ 0 & U \end{pmatrix}$ is negative definite on the column space of $\bar{G}$ and the rows of $\bar{H}$ are a basis for the null space of $\bar{G}^T$, then (12) is uniquely solvable for $\begin{pmatrix} x \\ \hline y \end{pmatrix}$ in terms of the problem parameters and data.

Lemma S will be applied in Section IV to derive convergence criteria for an iterative solution method for nonlinear versions of the spatial equilibrium problem. Lemma S is also used in an iterative method of sensitivity analysis, which is discussed in Section V.

IV. ITERATION

We focus on the spatial equilibrium conditions (1-6) as the fundamental problem to be solved. In the integrable case, the equivalent maximization problem (7) may be formulated and solved to obtain an equilibrium solution. In case the supply or demand function is *not* a gradient, i.e., the nonintegrable case, we consider the following iterative process to solve for equilibrium values of x and y. Assume that $t_{ij}(z) = t_{ij}$ for all $1 \leq i \leq m$, $1 \leq j \leq n$.

Step 1. Let $\begin{pmatrix} x^1 \\ \hline y^1 \end{pmatrix}$ be an initial guess and let $k = 1$. If desired, compute z^k, γ^k, and λ^k from

$$\begin{pmatrix} \bar{G}_x \\ \hline \bar{G}_y \end{pmatrix} \bar{z}^k = \begin{pmatrix} x^k \\ \hline y^k \end{pmatrix}$$

$\gamma^k = P_S(x^k)$, and $\lambda^k = P_D(y^k)$.

Step 2. Let $P_S^k(x) = P_S(x^k) + S^k(x - x^k)$ and $P_D^k(y) = P_D(y^k) + U^k(y - y^k)$, where S^k and U^k are *symmetric approximations* to $P_S'(x^k)$ and $P_D'(y^k)$, respectively, i.e., S^k is a symmetric $m \times m$ matrix and U^k is a symmetric $n \times n$ matrix. Note that P_S^k and P_D^k are linear approximations to P_S and P_D at x^k and y^k, respectively. Also, since S^k and U^k are symmetric matrices, there are quadratic functions g^k and f^k so that $\nabla g^k = P_S^k$ and $\nabla f^k = P_D^k$.

Step 3. Obtain $\begin{pmatrix} x^{k+1} \\ \hline y^{k+1} \end{pmatrix}$ as a solution of the quadratic programming problem

$$\max_{x,y,z} \; \{g^k(y) - f^k(x) - t^T z\}$$

$$\text{s.t.} \quad G_x z \leqq x$$

$$G_y z \geqq y$$

$$x, y, z \geqq 0$$

Step 4. Let

$$R_k = C_1 ||x^{k+1} - x^k|| + C_2 ||y^{k+1} - y^k||$$
$$+ C_3 ||-P_S(x^{k+1})^T \bar{G}_x + P_D(y^{k+1})^T \bar{G}_x - \bar{t}||$$
$$+ C_4 \left|\left|\bar{H}\begin{pmatrix} x^{k+1} \\ \hline y^{k+1} \end{pmatrix}\right|\right|$$

for preassigned nonnegative numbers C_i. If $R_k < \varepsilon$, terminate; otherwise, let $k \leftarrow k + 1$ and return to Step 2.

The main point of departure of this algorithm from that of the linear case is, of course, in Step 2, where $P_S'(x^k)$ and $P_D'(y^k)$ are reapproximated on each iteration, i.e., in general, S^k and U^k change between iterations in the nonlinear case.

The quantity $S^{k-1}(x^k - x^{k-1})$ represents the difference $P_S^{k-1}(x^k) - P_S^{k-1}(x^{k-1})$, which approximates $P_S(x^k) - P_S(x^{k-1})$. (Recall that $P_S^{k-1}(x^{k-1}) = P_S(x^{k-1})$.) If we let T^{k-1} denote an $m \times m$ matrix so that

$$T^{k-1}(x^k - x^{k-1}) = P_S(x^k) - P_S^{k-1}(x^k)$$

then T^{k-1} can be interpreted as the "bad part" of a secant approximation to P_S at x^{k-1}. This interpretation is illustrated in Figure 1 and by the following calculation:

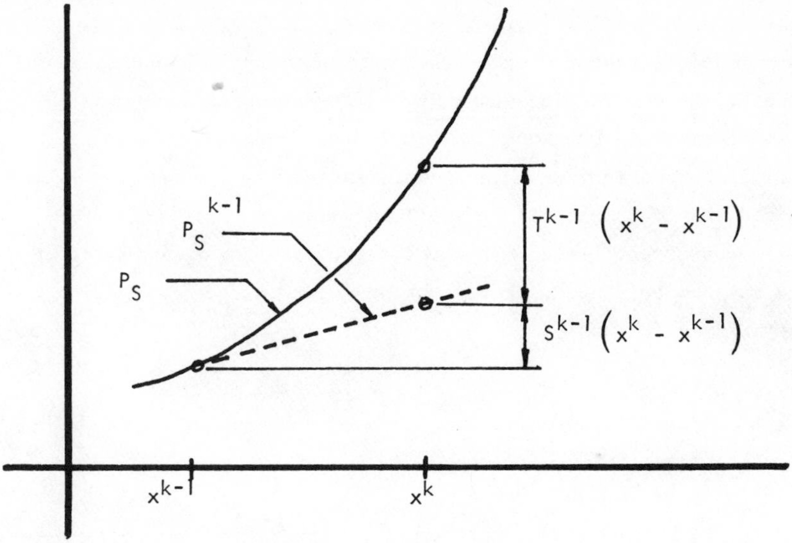

Figure 1 Illustration of Equation (13).

$$\begin{aligned} P_S(x^k) - P_S(x^{k-1}) &= [P_S(x^k) - P_S^{k-1}(x^k)] + [P_S^{k-1}(x^k) - P_S^{k-1}(x^{k-1})] \\ &= T^{k-1}(x^k - x^{k-1}) + S^{k-1}(x^k - x^{k-1}) \\ &= (S^{k-1} + T^{k-1})(x^k - x^{k-1}) \end{aligned} \tag{13}$$

Thus S^{k-1} and T^{k-1} are a splitting of a secant transformation determined by $(x^{k-1}, P_S(x^{k-1}))$ and $(x^k, P_S(x^k))$.

Similar remarks apply to the linear approximation of P_D at y^{k-1}. We let V^{k-1} denote an n × n matrix so that

$$P_D(y^k) - P_D^{k-1}(y^k) = V^{k-1}(y^k - y^{k-1})$$

and then obtain

$$P_D(y^k) - P_D(y^{k-1}) = (U^{k-1} + V^{k-1})(y^k - y^{k-1}) \tag{14}$$

which shows that U^{k-1} and V^{k-1} are a splitting of a secant transformation determined by $(y^{k-1}, P_D(y^{k-1}))$ and $(y^k, P_D(y^k))$.

The particular choice or construction of S^k and U^k in Step 2 is a matter for further analysis. We have used the most obvious choices such as

$U^k = \text{diag } P_D'(y^k)$ or $U^k = (1/2)[P_D'(y^k)^T + P_D'(y^k)]$ for log linear P_D in a supply/demand model of Eastern coal markets. Equations (13) and (14) suggest the possibility of adapting quasi-Newton, secant update methods in Step 2 of the iteration. The works of Dennis and Schnabel [3] on symmetric, positive definite approximations, and Dennis and Moré [2] on quasi-Newton methods, should provide a variety of implementations of Step 2.

A general convergence theorem for the iterative process outlined in Steps 1 to 4 may now be stated. For notational convenience, let A_k, B_k, and Δ_k be defined as follows:

$$A_k = \left(\begin{array}{c} \left(\bar{G}_x^T \mid \bar{G}_y^T\right)\begin{pmatrix} -S^k & 0 \\ 0 & U^k \end{pmatrix} \\ \hline \bar{H} \end{array}\right)$$

$$B_k = \left(\begin{array}{c} \left(\bar{G}_x^T \mid \bar{G}_y^T\right)\begin{pmatrix} T^k & 0 \\ 0 & -V^k \end{pmatrix} \\ \hline 0 \end{array}\right)$$

$$\Delta_k = \left(\begin{array}{c} x^{k+1} - x^k \\ \hline y^{k+1} - y^k \end{array}\right)$$

The following corollary to Lemma 2 establishes our iteration formula.

COROLLARY 1 If z^k and z^{k+1} have the same components positive, then $A_k\Delta_k = B_{k-1}\Delta_{k-1}$.

PROOF. Since $x^k, y^k, z^k, \gamma^k, \lambda^k$ is a solution of the spatial equilibrium conditions (1-6) with P_S and P_D replaced by P_S^{k-1} and P_D^{k-1}, Lemma 2 implies that

$$-P_S^{k-1}(x^k)^T\,\bar{G}_y + P_D^{k-1}(y^k)\,\bar{G}_y = \bar{t}^T$$

Under the assumption that z^{k+1} and z^k have the same components positive, i.e., $\bar{G}$ does not change from iteration k to iteration k + 1, we can obtain

$$-\bar{G}_x^T[P_S^k(x^{k+1}) - P_S^{k-1}(x^k)] + \bar{G}_x^T[P_D^k(y^{k+1}) - P_D^{k-1}(y^k)] = 0 \qquad (15)$$

Introducing S^k, T^{k-1}, U^k, and V^{k-1} into (15) yields

$$\left(\begin{array}{c} \left(\bar{G}_x^T \mid \bar{G}_y^T\right)\begin{pmatrix} -S^k & 0 \\ 0 & U^k \end{pmatrix} \\ \hline \bar{H} \end{array}\right)\left(\begin{array}{c} x^{k+1} - x^k \\ \hline y^{k+1} - y^k \end{array}\right) = \left(\begin{array}{c} \left(\bar{G}_x^T \mid \bar{G}_y^T\right)\begin{pmatrix} T^{k-1} & 0 \\ 0 & -V^{k-1} \end{pmatrix} \\ \hline 0 \end{array}\right)\left(\begin{array}{c} x^k - x^{k-1} \\ \hline y^k - y^{k-1} \end{array}\right)$$

which, with obvious notational changes, establishes the result of the corollary,

$$A_k \Delta_k = B_{k-1} \Delta_{k-1} \tag{16}$$

An application of Lemma S to equation (16) yields the following corollary. Assume that the assumptions of Lemma S are satisfied.

COROLLARY 2 If, in addition to being symmetric, S^k and U^k are chosen so that $\begin{pmatrix} -S^k & 0 \\ 0 & U^k \end{pmatrix}$ is negative definite on the column space of $\bar{G} = \begin{pmatrix} \bar{G}_x \\ \bar{G}_y \end{pmatrix}$, then A_k is invertible and (16) is uniquely solvable for Δ_k.

The first convergence result may now be stated.

CONVERGENCE RESULT 1. Suppose P_S and P_D are continuous and there is a positive integer, K, so that (i) for some norm, $||\cdot||$, on $R^{(m+n)\times(m+n)}$, $||A_k^{-1} B_{k-1}|| \leqq r < 1$ for all $k > K$, and (ii) all z^k for $k > K$ have the same components positive, i.e., $\bar{G}$ does not change for $k > K$; then the sequences $\{x^k\}_{k=1}^{\infty}$ and $\{y^k\}_{k=1}^{\infty}$ generated by Steps 1 - 4 converge to a solution of the spatial equilibrium conditions (1-6).

INDICATION OF PROOF. Parts (i) and (ii) of the hypothesis easily imply, via a contraction type of argument, that the sequences $\{x^k\}_{k=1}^{\infty}$ and $\{y^k\}_{k=1}^{\infty}$ converge. Letting x and y denote the sequential limits (in an actual computation, $x = x^k$ and $y = y^k$ for some positive integer k), z, γ, and λ may be determined from $\bar{G}z = \begin{pmatrix} x \\ y \end{pmatrix}$, $\gamma = P_S(x)$, and $\lambda = P_D(y)$. The continuity of P_S and P_D and the form of P_S^k and P_D^k enable one to conclude that x, y, z, γ, λ is a solution of (1-6).

We will now prove convergence of the sequences $\{x^k\}_{k=1}^{\infty}$ and $\{y^k\}_{k=1}^{\infty}$ in a way that does not make assumptions such as condition (ii) in Convergence Result 1; i.e., we will make no assumptions about which components of z^k are positive. The approach here is to relate the equilibrium conditions (1-6) to a nonlinear complementarity problem and then apply a convergence theorem that the first author presented at the 1980 SIAM fall meeting.

It is well known that the nonlinear complementarity problem

$$\begin{aligned} & w \geqq 0,\ F(w) \geqq 0 \\ & F(w)^T w = 0 \end{aligned} \tag{17}$$

is a restatement of (1-6) with

$$w = \begin{pmatrix} x \\ y \\ z \\ \gamma \\ \lambda \end{pmatrix} \quad \text{and} \quad F(w) = \begin{pmatrix} P_S(x) - \gamma \\ \lambda - P_D(y) \\ G_x^T\gamma + t(z) - G_y^T\lambda \\ x - G_x z \\ G_y z - y \end{pmatrix}$$

It is also true, that if $\{w^k\}_{k=1}^{\infty}$ is a sequence so that for each k

$$\begin{aligned} &w^{k+1} \geqq 0, \; F^k(w^{k+1}) \geqq 0 \\ &F^k(w^{k+1})^T w^{k+1} = 0 \end{aligned} \tag{18}$$

then

$$\langle w^{k+1} - w^k, F^k(w^{k+1}) - F^{k-1}(w^k)\rangle \leqq 0 \tag{19}$$

Notation: Let $\langle U,V\rangle$ denote the inner product of two vectors U and V.

LEMMA 4 Let $\{w^k\}_{k=1}^{\infty}$ be a sequence satisfying (19), where F^k is defined by

$$F^k(w) = F(w^k) + S^k(w - w^k) \tag{20}$$

Also, let T^k be a matrix so that

$$T^k(w^{k+1} - w^k) = F(w^{k+1}) - F(w^k) - S^k(w^{k+1} - w^k) \tag{21}$$

then

$$\langle\Delta^k, S^k\Delta^k\rangle + \langle\Delta^k, T^{k-1}\Delta^{k-1}\rangle \leqq 0 \tag{22}$$

where Δ^k denotes $w^{k+1} - w^k$.

PROOF. Using the definition of F^k and equation (21), we obtain

$$\begin{aligned} F^k(w^{k+1}) - F^{k-1}(w^k) &= F(w^k) + S^k(w^{k+1} - w^k) - F(w^{k-1}) - S^{k-1}(w^k - w^{k-1}) \\ &= S^k\Delta^k + T^{k-1}\Delta^{k-1} \end{aligned}$$

which can be substituted into (19) to obtain the results of the lemma.

Recall that x^{k+1}, y^{k+1}, z^{k+1}, γ^{k+1}, λ^{k+1}, as generated by the four-step iterative process described earlier in this section, is a solution of the equilibrium conditions (1-6) with linear t and with P_S and P_D replaced by P_S^k and P_D^k. Therefore, if we let

$$w^k = \begin{pmatrix} x^k \\ y^k \\ z^k \\ \gamma^k \\ \lambda^k \end{pmatrix}$$

then w^{k+1} satisfies (18), where F^k is defined by (20) and

$$\mathcal{S}^k = \begin{pmatrix} S^k & 0 & 0 & -I & 0 \\ 0 & -U^k & 0 & 0 & I \\ 0 & 0 & 0 & G_x^T & -G_y^T \\ I & 0 & -G_x & 0 & 0 \\ 0 & -I & G_y & 0 & 0 \end{pmatrix} \tag{23}$$

With $\mathcal{S}^k$ defined by (23), it is straightforward to show that $\mathcal{T}^k$ is given by

$$\mathcal{T}^k = \begin{pmatrix} T^k & 0 & 0 & 0 & 0 \\ 0 & -V^k & 0 & 0 & 0 \\ 0 & 0 & 0 & 0 & 0 \\ 0 & 0 & 0 & 0 & 0 \\ 0 & 0 & 0 & 0 & 0 \end{pmatrix} \tag{24}$$

With $\mathcal{S}^k$ and $\mathcal{T}^k$ given by (23) and (24), we can substitute into (22) and obtain that

$$\left\langle \begin{pmatrix} x^{k+1} - x^k \\ y^{k+1} - y^k \end{pmatrix}, \begin{pmatrix} S^k & 0 \\ 0 & -U^k \end{pmatrix} \begin{pmatrix} x^{k+1} - x^k \\ y^{k+1} - y^k \end{pmatrix} + \begin{pmatrix} T^{k-1} & 0 \\ 0 & -V^{k-1} \end{pmatrix} \begin{pmatrix} x^k - x^{k-1} \\ y^k - y^{k-1} \end{pmatrix} \right\rangle \leqq 0 \tag{25}$$

Inequality (25) is an important part of the following convergence theorem.

THEOREM C Suppose S^k and $-U^k$ are symmetric and positive definite, inequality (25) holds, and there is a positive integer K and number M so that for all $k > K$,

(i) $\left\|\sqrt{S^k}^{\,-1}\right\| + \left\|\sqrt{-U^k}^{\,-1}\right\| \leqq M$, and

(ii) $\left\|\sqrt{S^k}^{\,-1} T^{k-1} \sqrt{S^{k-1}}^{\,-1}\right\| + \left\|\sqrt{-U^k}^{\,-1} V^{k-1} \sqrt{-U^{k-1}}^{\,-1}\right\| \leqq r < 1$, then $\{x^k\}_{k=1}^{\infty}$ and $\{y^k\}_{k=1}^{\infty}$ converge.

INDICATION OF PROOF. For notational convenience, let

$$\delta^k = \begin{pmatrix} x^{k+1} - x^k \\ y^{k+1} - y^k \end{pmatrix}$$

$$\Sigma^k = \begin{pmatrix} S^k & 0 \\ 0 & -U^k \end{pmatrix}$$

and

$$\Gamma^k = \begin{pmatrix} T^k & 0 \\ 0 & -V^k \end{pmatrix}$$

From (25) we have

$$\langle \delta^k, \Sigma^k \delta^k \rangle + \langle \delta^k, \Gamma^{k-1} \delta^{k-1} \rangle \leqq 0$$

Since S^k and $-U^k$ are symmetric and positive definite, $\sqrt{\Sigma^k}$ and $\sqrt{\Sigma^k}^{-1}$ exist and so we have

$$\left\langle \sqrt{\Sigma^k}\, \delta^k, \sqrt{\Sigma^k}\, \delta^k \right\rangle + \left\langle \sqrt{\Sigma^k}\, \delta^k, \sqrt{\Sigma^k}^{-1}\, \Gamma^{k-1} \delta^{k-1} \right\rangle \leqq 0$$

for all k. Using standard properties of norms and inner products, we obtain

$$||\sqrt{\Sigma^k}\, \delta^k||^2 \leqq ||\sqrt{\Sigma^k}\, \delta^k||\ ||\sqrt{\Sigma^k}^{-1}\, \Gamma^{k-1} \sqrt{\Sigma^{k-1}}||\ ||\sqrt{\Sigma^{k-1}}\, \delta^{k-1}||$$

which can be simplified to

$$||\sqrt{\Sigma^k}\, \delta^k|| \leqq \alpha_k ||\sqrt{\Sigma^{k-1}}\, \delta^{k-1}||$$

where

$$\alpha_k = ||\sqrt{\Sigma^k}^{-1}\, \Gamma^{k-1} \sqrt{\Sigma^{k-1}}^{-1}||$$

Substituting for Σ^k, Γ^k, Σ^{k-1} we obtain

$$\alpha_k = ||\sqrt{\Sigma^k}^{-1}\, \Gamma^{k-1} \sqrt{\Sigma^{k-1}}^{-1}||$$

$$= \left|\left| \begin{pmatrix} \sqrt{S^k}^{-1} & 0 \\ 0 & \sqrt{-U^k}^{-1} \end{pmatrix} \begin{pmatrix} T^{k-1} & 0 \\ 0 & -V^{k-1} \end{pmatrix} \begin{pmatrix} \sqrt{S^{k-1}}^{-1} & 0 \\ 0 & \sqrt{-U^{k-1}}^{-1} \end{pmatrix} \right|\right|$$

$$= \left\| \begin{pmatrix} \sqrt{S^k}^{-1} T^{k-1} \sqrt{S^{k-1}}^{-1} & 0 \\ 0 & \sqrt{-U^k}^{-1} (-V^{k-1}) \sqrt{-U^{k-1}}^{-1} \end{pmatrix} \right\|$$

$$\leqq \left\| \sqrt{S^k}^{-1} T^{k-1} \sqrt{S^{k-1}}^{-1} \right\| + \left\| \sqrt{-U^k}^{-1} V^{k-1} \sqrt{-U^{k-1}}^{-1} \right\|$$

Condition (ii) in the hypothesis implies that $\alpha_k \leqq r < 1$ for all k and condition (i) implies that $\{\|\sqrt{\Sigma^k}^{-1}\|\}_{k=1}^{\infty}$ are uniformly bounded by M; therefore the sequences $\{x^k\}_{k=1}^{\infty}$ and $\{y^k\}_{k=1}^{\infty}$ converge.

Let x and y denote the sequential limits and find z, γ, and λ so that $G_x z = x$, $G_y z = y$, $\gamma = P_S(x)$, and $\lambda = P_D(y)$. Again, as in the proof of Convergence Result 1, continuity of P_S and P_D and the form of P_S^k and P_D^k enable one to conclude that x, y, z, γ, λ is a solution of the equilibrium conditions (1-6).

The hypothesis of continuity is strong; however, it was satisfied by the linear P_S/log linear P_D functions in our model of Eastern coal markets.

Certainly, the objective of much research has been to compute equilibria with far less restrictive conditions on the supply and demand relations; however, our main point here is to observe the common hypotheses for iteration convergence and sensitivity analysis in the more restrictive case. Ahn's dissertation [1] contains convergence theorems and applications of complementarity methods in energy modeling. In [10], Pang presents several general theorems, classified according to method of proof, for convergence of iterative techniques in complementarity and variational problems.

V. SENSITIVITY

Now, suppose that a solution x, y, z, γ, λ of the spatial equilibrium conditions (1-6) has been obtained. The questions of "how much" and "in what direction" the equilibrium solution changes for small perturbations in the problem parameters are the classic questions of local sensitivity analysis. Our approach to sensitivity is motivated by Chapter 2 of Fiacco and McCormick [4]. In this section we shall show that the local sensitivity of an equilibrium solution can be obtained from information that is already available from the iterative solution method discussed in Section IV.

Again, assume that the transportation costs satisfy $t_{ij}(z) = \bar{t}_{ij}$. Let us suppose that some parameters in P_S, P_D, or $\bar{t}$ are perturbed by an amount

ε, and let $x(\varepsilon)$, $y(\varepsilon)$, $z(\varepsilon)$, $\gamma(\varepsilon)$, and $\lambda(\varepsilon)$, denote a solution of the perturbed problem. As a result of Lemmas 1 and 2 the following equations hold:

$$-\bar{G}_x^T P_S(x(\varepsilon),\varepsilon) + \bar{G}_y^T P_D(y(\varepsilon),\varepsilon) = \bar{t}(\varepsilon) \tag{26}$$

$$\bar{H}\begin{pmatrix} x(\varepsilon) \\ \hline y(\varepsilon) \end{pmatrix} = 0 \tag{27}$$

If we assume that (26) and (27) determine x and y as differentiable functions of ε, then we can obtain

$$\begin{bmatrix} \left(\bar{G}_x^T \mid \bar{G}_y^T\right)\begin{pmatrix} -\dfrac{\partial P_S}{\partial x}(x(0),0) & 0 \\ 0 & \dfrac{\partial P_D}{\partial y}(y(0),0) \end{pmatrix} \\ \hline \bar{H} \end{bmatrix}\begin{pmatrix} x'(0) \\ \hline y'(0) \end{pmatrix} = \begin{bmatrix} \left(\bar{G}_x^T \mid \bar{G}_y^T\right)\begin{pmatrix} \dfrac{\partial P_S}{\partial \varepsilon}(x(0),0) \\ \hline -\dfrac{\partial P_D}{\partial \varepsilon}(x(0),0) \end{pmatrix} + \dfrac{\partial \bar{t}}{\partial \varepsilon} \\ \hline 0 \end{bmatrix} \tag{28}$$

where $\partial P_S/\partial x$ denotes the $m \times m$ matrix function whose ith row is $(\nabla_x P_S^T)_i$, and $\partial P_D/\partial y$ is an $n \times n$ matrix function whose jth row is $(\nabla_y P_D^T)_j$,

$$x'(\varepsilon) = (x_1'(\varepsilon), \ldots, x_m'(\varepsilon))^T$$

and

$$y'(\varepsilon) = (y_1'(\varepsilon), \ldots, y_n'(\varepsilon))^T$$

The components of the $(m + n)$ vector $\begin{pmatrix} x'(0) \\ \hline y'(0) \end{pmatrix}$ in (28) are the local sensitivities of x and y with respect to the particular parameters that were perturbed. We propose to solve for the sensitivity numbers by letting S and U be symmetric approximations to $(\partial P_S/\partial x)(x(0),0)$ and $(\partial P_D/\partial y)(y(0),0)$, respectively. Then (28) can be rewritten as

$$\begin{bmatrix} \left(\bar{G}_x^T \mid \bar{G}_y^T\right)\begin{pmatrix} -S & 0 \\ 0 & U \end{pmatrix} \\ \hline \bar{H} \end{bmatrix}\begin{pmatrix} x'(0) \\ \hline y'(0) \end{pmatrix} = \begin{bmatrix} \left(\bar{G}_x^T \mid \bar{G}_y^T\right)\begin{pmatrix} T & 0 \\ 0 & -V \end{pmatrix} \\ \hline 0 \end{bmatrix}\begin{pmatrix} x'(0) \\ \hline y'(0) \end{pmatrix} + \begin{bmatrix} \left(\bar{G}_x^T \mid \bar{G}_y^T\right)\begin{pmatrix} \partial P_S/\partial \varepsilon \\ \hline \partial P_D/\partial \varepsilon \end{pmatrix} + \dfrac{\partial \bar{t}}{\partial \varepsilon} \\ \hline 0 \end{bmatrix} \tag{29}$$

where

$$T = \frac{\partial P_S}{\partial x}(x(0),0) - S \qquad \text{and} \qquad V = \frac{\partial P_D}{\partial y}(y(0),0) - U$$

With obvious notational changes, (29) is a system of equations of the form

$$Aw = Bw + C \tag{30}$$

The form of (30) suggests an iterative method to solve for $w = \begin{pmatrix} x'(0) \\ \hline y'(0) \end{pmatrix}$ via

$$Aw_{k+1} = Bw_k + C \tag{31}$$

Lemma S applies here, as it did in Section IV, to yield the following solvability result.

COROLLARY 3 If, in addition to being symmetric, S and U are chosen so that $\begin{pmatrix} -S & 0 \\ 0 & U \end{pmatrix}$ is negative definite on the column space of $\bar{G} = \begin{pmatrix} G_x \\ \hline -G_y \end{pmatrix}$, then (29) is uniquely solvable for $\begin{pmatrix} x'(0) \\ \hline y'(0) \end{pmatrix}$.

Again, paralleling the development for iteration convergence, we have the following theorem.

THEOREM S If the spectral radius of $A^{-1}B$, $\rho(A^{-1}B)$, is less than one, then (31) converges geometrically to a solution $\begin{pmatrix} x'(0) \\ \hline y'(0) \end{pmatrix}$ of (28).

For a proof of Theorem S see references on iterative methods such as Chapter 7 of Strang [12], Chapter 3 of Varga [14], or Chapters 5 and 12 of Ortega and Rheinboldt [9].

In case the equilibrium solution is determined by the iterative method discussed in Section IV, natural candidates for S and U in (29) would be S^k and U^k from Step 2 of the iterative process, where k is the positive integer so that $x = x^k$ and $y = y^k$. We would then have $S \simeq S^k$ and $U \simeq U^k$, and so the convergence condition $||A_k^{-1}B_k|| < 1$ from Convergence Result 1 will imply the condition $\rho(A^{-1}B) < 1$ in Theorem S provided that $T - T^k$ and $V - V^k$ are sufficiently small. A more rigorous treatment of the relationship between these conditions will be taken up in future work.

VI. CONCLUSION

Our original interest in this problem was simply to know when equation (28) was uniquely solvable for the sensitivity numbers $\begin{pmatrix} x'(0) \\ \hline y'(0) \end{pmatrix}$. At the same time, we knew that our model would be much better if cross price effects in P_D were not necessarily equal. This suggested an iterative solution method, similar to that which was used by Hogan [5] in the PIES energy model equilibrating algorithm. Expecting similarities in sensitivity analysis and algorithm convergence, as noted in the introduction of this paper, we were led to the solvability lemma of Section III.

The main purpose, then of this paper is to observe the close link between sensitivity analysis and iterative methods for nonintegrable, spatial equilibrium models. This linkage is through Lemma N, Lemma S, and the conditions for convergence as stated in the hypothesis of Convergence Result 1 and Theorem S.

We suggest further that in general, many iterative solution methods for nonlinear problems can be readily extended to include an analysis of the sensitivity of the solution vector to perturbations of the problem parameters. The reason is that iterative methods as well as sensitivity analysis can be based on local (perhaps linear) approximations to the underlying nonlinear functions; therefore, the information needed to carry out a sensitivity analysis is already available in the iterative method.

VII. ACKNOWLEDGMENTS

The first author gratefully acknowledges a conversation with J. S. Pang that motivated inequality (19). The cooperation and helpful suggestions of the editor of these proceedings are very much appreciated, as are the time and constructive comments of the referees.

REFERENCES

1. B. H. Ahn. Computation of market equilibria for policy analysis: the Project Independence Evaluation System approach. Ph.D. dissertation, Stanford University, 1978.

2. J. E. Dennis and J. J. Moré. Quasi-Newton methods, motivation and theory, *SIAM Review*, Vol. 19, pp. 46-89 (1977).

3. J. E. Dennis and R. B. Schnabel. A new derivation of symmetric positive definite secant updates, Technical Report CU-CS-185-80, Department of Mathematical Sciences, Rice University, Houston, Texas (1980).

4. A. V. Fiacco and G. P. McCormick. *Nonlinear Programming: Sequential Unconstrained Minimization Techniques*. Wiley, New York, 1968.

5. W. W. Hogan. Project Independence Evaluation System: structure and algorithms, *Proceedings of Symposia in Applied Mathematics*, Vol. 21, American Mathematical Society (1976).

6. C. L. Irwin and C. W. Yang. Iteration and sensitivity for a spatial equilibrium problem with linear supply and demand functions, *Operations Research*, Vol. 30, pp. 319-335 (1982).

for publication in *Operations Research*.

7. M. Kennedy. An economic model of the world oil market, *Bell Journal of Economics and Management Science*, Vol. 5 (1974).

8. G. P. McCormick. Optimality criteria in nonlinear programming, *Proceedings of a Symposium in Applied Mathematics* (R. W. Cottle and C. E. Lemke, eds.), Vol. 9, SIAM-AMS Proceedings (1975).

9. J. M. Ortega and W. C. Rheinboldt. *Iterative Solution of Nonlinear Equations in Several Variables*. Academic Press, New York, 1970.

10. J. S. Pang and D. Chan. Iterative methods for variational and complementarity problems, Preprint, Graduate School of Industrial Administration, Carnegie-Mellon University (1981).

11. E. Silberberg. A theory of spatially separated markets, *International Economic Review* (1977).

12. G. Strang. *Linear Algebra and Its Applications*. Academic Press, New York, 1976.

13. T. Takayama and G. G. Judge. *Spatial and Temporal Price and Allocation Methods*. North Holland, Amsterdam, 1971.

14. R. S. Varga. *Matrix Iterative Analysis*. Prentice-Hall, Englewood Cliffs, New Jersey, 1962.

CHAPTER 7 A SENSITIVITY ANALYSIS APPROACH TO ITERATION SKIPPING IN THE HARMONIC MEAN ALGORITHM*

JOHN J. DINKEL / Texas A&M University, College Station, Texas

G. A. KOCHENBERGER / The Pennsylvania State University, University Park, Pennsylvania

DANNY S. WONG / Ohio State University, Columbus, Ohio

ABSTRACT

This paper explores acceleration of the harmonic mean algorithm for polynomial geometric programming using sensitivity analysis procedures. This acceleration is accomplished by using the sensitivity analysis methods, developed for ordinary geometric programs, to skip iterations of the harmonic algorithm. Several implementations of the acceleration procedure based on the size of the sensitivity analysis steps, frequency of resolving the harmonic program, and use of incremental procedure as a means of controlling step size are described and tested on a set of problems. The comparison of the implementations is based on the criteria of computational expense and accuracy of the solutions.

I. INTRODUCTION

The motivation for this paper is an observation made by Duffin and Peterson [11] in the context of their development of the harmonic mean approach to polynomial geometric programs. During the early to mid-1970s there were several

---

*Revised July 31, 1981.

developments that led to numerical approaches to solving this more general class of programs [3,11,12]. The essence of these methods is to deal with the polynomial terms by "condensing" them to a single term which then can be dealt with using posynomial geometric programming techniques. Among these developments are the "geometric inequalities" of Duffin and Peterson [12], in particular, the harmonic mean approach [11]. As will be described in detail in the next section, one approach has the property of maintaining the invariance of the matrix of exponents associated with the primal program with regard to the approximation of the nonposynomial terms, a property not shared by the other approaches. As Duffin and Peterson [11] observed, one should be able to exploit this property to improve numerical solution procedures.

When the observation above was made, there was very little in the way of implementable approaches to either the harmonic mean algorithm or sensitivity analysis procedures. The purpose of this paper is to utilize some of the more recent implementations, particularly with regard to sensitivity analysis in geometric programming, to study the acceleration of the harmonic mean algorithm.

The original characterization of the effect of perturbations on the optimal solution was given by Duffin, Peterson, and Zener [13], and was restricted to the coefficients of the objective function and constraints. The duality theory of geometric programming plays a central role in these results, for the changes in the primal parameters are studied via their impact on the dual geometric program. These results have been extended and implemented by Dinkel, *et al*. [5,6,8,9], and provide a usable set of procedures for performing such analyses. The seminal work of Fiacco and McCormick [15] and the recent paper by Bigelow and Shapiro [4] provide an approach to studying the effect of any parameter changes, including the primal exponents. In addition, the work of Armacost and Fiacco [1,2] provides an alternative primal method for sensitivity analysis. This paper brings these methods together in the study of accelerating the harmonic mean algorithm using sensitivity analysis methods.

II. POLYNOMIAL GEOMETRIC PROGRAMMING

Ordinary geometric programming, as developed in Duffin, *et al*. [13], consists of a *primal program* A:

$$\begin{aligned} &\text{minimize} \quad g_0(t) \\ &\text{subject to } g_k(t) \leqq 1 \qquad k = 1,\ldots,p \\ &\qquad\qquad t > 0 \end{aligned} \tag{1}$$

where

$$g_k(t) = \sum_{i=m_k}^{n_k} c_i \prod_{j=1}^{m} t_j^{a_{ij}} \qquad k = 0,1,\ldots,p$$

and

$$c_i \geqq 0$$

The indices m_k, n_k are used to number the terms in the primal program consecutively; for example, $n = 1,\ldots,n_0$ for the objective function, $m_1 = n_0 + 1,\ldots,n_1$ for the first constraint, and so on, with $n_p = n$.

The associated dual geometric program is B:

$$\text{maximize} \quad v(\delta) = \prod_{i=1}^{n} (c_i/\delta_i)^{\delta_i} \prod_{k=1}^{p} \lambda_k^{\lambda_k}$$

$$\begin{aligned} &\text{subject to } \sum_{i=1}^{n_0} \delta_i = 1 \\ &\sum_{i=1}^{n} a_{ij}\delta_i = 0 \qquad j = 1,\ldots,m \\ &\delta_i \geqq 0 \qquad i = 1,\ldots,n \end{aligned} \tag{2}$$

where

$$\lambda_k = \sum_{i=m_k}^{n_k} \delta_i \qquad k = 1,\ldots,p$$

As noted in [13], ordinary geometric programming is a subset of convex programming, since each program of the form (1) can be transformed into a convex program.

Thus, to consider a wider range of problems it is necessary to relax the assumption that $c_i \geqq 0$. This relaxation allows us to consider primal programs where, among other things, the optimization operator is maximize and constraints are of the form $g_k(t) \geqq 1$. Also, as shown by Duffin and Peterson [11,12], this allows for the consideration of algebraic programs as well.

Following the development of Duffin and Peterson [11,12], we assume the original polynomial problem has been written as a *reversed geometric program*, A_R:

$$\begin{aligned} &\text{minimize} \quad g_0(t) \\ &\text{subject to} \quad g_k(t) \leqq 1 \qquad k = 1,\ldots,p \\ &\qquad\qquad\quad g_k(t) \geqq 1 \qquad k = p+1,\ldots,r \\ &\qquad\qquad\quad t > 0 \end{aligned} \tag{3}$$

where all $g_k(t)$ are posynomials. The constraints $g_k(t) \geqq 1$, $k = p + 1,\ldots,r$, are referred to as reversed constraints; hence the name for the family of programs. The details of the necessary transformations to obtain A_R are given in [12].

There have been a number of approaches to the solution of polynomial geometric programs based on the solution of sequences of ordinary geometric programs. These methods are based on "condensing" the nonposynomial terms, for example, $g_k(t) \geqq 1$ in (3), to single term posynomials which then can be handled using ordinary geometric programming algorithms. The strategies for this condensation include the complementary program of Avriel and Williams [3], the geometric-mean programs of Duffin and Peterson [11,12], and the linearization or cutting plane approaches [10].

The bases for the various approaches are the classical inequalities given in [12, Lemma 3a] as

$$\left(\sum_{i=1}^{n} u_i\right)^{-1} \leqq \prod_{i=1}^{n} \left(\frac{\alpha_i}{u_i}\right)^{\alpha_i} \leqq \sum_{i=1}^{n} \frac{\alpha_i^2}{u_i}$$

where $u_1,\ldots,u_n$ are positive and $\alpha_i \geqq 0$ and $\sum_{i=1}^{n} \alpha_i = 1$. Thus, given a posynomial $g(t) = \sum_{i=1}^{n} u_i(t)$ and weights α_i as above, we can define approximating geometric programs where constraints of the form $g_k(t) \geqq 1$ are replaced by prorotype constraints of the form $g_k(t;\alpha) \leqq 1$. Complete details of the various approaches can be found in [3,10,11,12].

For our purposes here we concentrate on the harmonic mean approach [11] which approximates A_R by the *harmonic program*, $A_H(\alpha)$:

$$\begin{aligned} &\text{minimize} \quad g_0(t) \\ &\text{subject to} \quad g_k(t) \leqq 1 \qquad k = 1,\ldots,p \\ &\qquad\qquad\quad g_k(t;\alpha) \leqq 1 \qquad k = p+1,\ldots,r \end{aligned} \tag{4}$$

where

$$g_k(t;\alpha) = \sum_{i=m_k}^{n_k} \alpha_i^2/u_i(t) \qquad k = p + 1,\ldots,r \tag{5}$$

and

$$\alpha_i = u_i(t)/g_k(t) \qquad \text{for any feasible } t \tag{6}$$

Since A_H is an ordinary geometric program, the associated dual geometric program, $B_H(\alpha)$, is:

$$\text{maximize} \quad v(\alpha) = \prod_{i=1}^{n_p} (c_i/\delta_i)^{\delta_i} \prod_{i=m_{p+1}}^{n} \left(\frac{\alpha_i^2/c_i}{\delta_i}\right)^{\delta_i} \prod_{k=1}^{r} \lambda_k^{\lambda_k} \tag{7}$$

$$\text{subject to} \quad \sum_{i=1}^{n_0} \delta_i = 1 \tag{8}$$

$$\sum_{i=1}^{n} a_{ij}\delta_i = 0 \qquad j = 1,\ldots,m \tag{9}$$

$$\delta_i \geqq 0 \qquad i = 1,\ldots,n \tag{10}$$

While the dual programs associated with the various condensation approaches have the same general structure, they differ in one important result:

> *For the harmonic program the system of dual constraints (8-10) is invariant with respect to the weights α_i. Also the objective function is separable (in log-linear form) with respect to the coefficients (α_i^2/c_i) of the reversed constraints. In the other condensation approaches, the primal exponents, a_{ij} in (9), become functions of the α_i. Similarly, the coefficients in the objective function represent the condensation of all coefficients of each constraint and hence are not separable in the α_i.*

While this invariance has been known for some time, it has not been exploited. In fact, in a comparison of the various condensation approaches [3,10,12] the harmonic approach consistently required more iterations and computational effort to obtain equivalent solutions. However, the exploitation of the invariance of the (a_{ij}) matrix with respect to α was not part of that study.

The solution strategy for polynomial geometric programs is based on the duality theory relating A_R, $A_H(\alpha)$, and $B_H(\alpha)$ as given in [11]. The algorithm is briefly described as:

Step 1: For a feasible solution to A_R, t^ℓ, construct the approximating program $A_H(\alpha^{\ell+1})$, where $\alpha^{\ell+1} = u_i(t^\ell)/g_k(t^\ell)$ and

$$\sum_{i=m_k}^{n_k} \alpha_i^{\ell+1} = 1 \qquad \text{for } k = p+1,\ldots,r$$

If the termination criterion is satisfied, the optimal solution is t; if not, go to Step 2.

Step 2: Determine the optimal solution to $A_H(\alpha^{\ell+1})$; call it $t^{\ell+1}$ and go back to Step 1.

While several termination criteria are possible, we adopt the criterion $|\alpha^{\ell+1} - \alpha^\ell| <$ TOLERANCE based on the result of [11], where TOLERANCE is given a predetermined value.

It is important to note that since A_R is not necessarily convex, the above algorithm may terminate at a local solution or at a stationary point. However, for each choice of weights, α^ℓ, the approximating program $A_H(\alpha^\ell)$ is convex and hence we can generate an optimal solution to the subproblem. The above algorithm is based on the following results from [11,12].

DEFINITION 1 A feasible solution t* to program A_R is called an equilibrium solution if there is a feasible solution δ* to the dual program B such that

$$\delta_i^* g_0(t^*) = u_i(t^*) \qquad i = 1,\ldots,n_0$$

$$\delta_i^* = \lambda_k^* u_i(t^*) \qquad i = m_1,\ldots,n$$

THEOREM 1 If t is a feasible solution to $A_H(\alpha)$, then t is feasible for A_R and hence $M_{A_H(\alpha)} \geq M_{A_R}$, where $M_{A_H(\alpha)}$, M_{A_R} are the minima of $A_H(\alpha)$ and A_R, respectively. On the other hand, if t is feasible for A_R, then t is a feasible solution to $A_H(\alpha)$ with $\alpha_i = u_i(t)/g_i(t)$, $i \in [k]$, $k \in R$.

THEOREM 2 For $t^* > 0$ define $\alpha_i^* = u_i(t^*)/g_k(t^*)$ for i[k], $k \in R$. Then t* and δ* are optimal solutions to $A_H(\alpha^*)$ and $B_H(\alpha^*)$, respectively, if and only if they are equilibrium solutions to A_R and B_R, in which case $M_{A_H(\alpha^*)} = M_{B_H(\alpha^*)}$, where B_R is the geometric programming dual to A_R and $M_{A_H(\alpha^*)}$, $M_{B_H(\alpha^*)}$ are the minima of $A_H(\alpha^*)$ and $B_H(\alpha^*)$, respectively.

III. SENSITIVITY ANALYSIS AND ALGORITHM ACCELERATION

Since each approximating program $A_H(\alpha^\ell)$ is a posynomial program, we propose a sensitivity analysis based modification of the harmonic mean algorithm. In particular, each iteration of the harmonic algorithm generated changes in certain parameters, specifically from $(\alpha_i^\ell)^2$ to $(\alpha_i^{\ell+1})^2$. These changes can be viewed as perturbations in the appropriate coefficients and sensitivity analysis procedures can be used to generate a new solution. In what follows in Step 2 of the harmonic mean algorithm, we replace the generation of an optimal solution to $A_H(\alpha^{\ell+1})$ by a sensitivity analysis generated solution.

That is, for a given α^ℓ the program $A_H(\alpha^\ell)$ is an ordinary geometric program. The harmonic mean algorithm would generate the next program in the sequence by determining an optimal solution, $t^{\ell+1}$, to $A_H(\alpha^\ell)$. This point is used to generate a new set of weights,

$$\alpha_i^{\ell+1} = u_i(t^{\ell+1})/g_k(t^{\ell+1}) \qquad i \in [k],\ k = m_1,\ldots,n$$

As can be seen in (5), changes in α_i correspond to changes in the appropriate coefficients. Thus it is possible to generate a new approximating optimal solution to $A_H(\alpha^{\ell+1})$ using the current optimal solution t^ℓ, the parameter changes from α^ℓ to $\alpha^{\ell+1}$, and the sensitivity analysis procedures of ordinary geometric programming.

In order to implement such a modification, we briefly review the appropriate sensitivity analysis procedures. For complete details and computational experience, the reader is referred to [5,6,7,8,9]. The sensitivity analysis procedures of Duffin, *et al.* [13] focus on the dual geometric program. That is, the effect of a parameter change on the primal program is studied via the dual geometric program. All calculations are aimed at determining a new dual solution from which the corresponding primal solution is computed using the duality theory of geometric programming [13]. Similarly, the results of Section 4 of Bigelow and Shapiro [4] are clearly applicable to the dual geometric program. Thus these two approaches provide us with alternate approaches to the implementation of the sensitivity analysis procedures.

A key point, no matter which method is used, is that for each α^ℓ the program $A_H(\alpha^\ell)$ is a posynomial geometric program. Previous results indicate the ability of sensitivity analysis procedures, particularly with incrementing, to generate feasible and accurate solutions. In the present context, (4-6), we will concentrate on changes in the constraints since the coefficients of the constraints $g_k(t;\alpha) \leq 1$, $k = p + 1,\ldots,r$, will be changing as the α_i change.

From the point of view of Duffin, *et al.* [13], changes in the α_i are reflected as changes dc_i for which the change in the dual solution is:

$$d\delta_i = \sum_{j=1}^{d} \left\{ b_i^{(j)} \sum_{\ell=1}^{d} \left[J_{j\ell}^{-1}(\delta^*) \sum_{i=1}^{n} b_i^{(\ell)} (dc_i/c_i) \right] \right\} \tag{11}$$

$$\frac{dv}{v^*} = \sum_{i=1}^{n} \delta_i^* dc_i/c_i \tag{12}$$

where

$$J_{j\ell}(\delta^*) = \sum_{q=1}^{n} b_q^{(j)} b_q^{(\ell)}/\delta_q^2 - \sum_{k=1}^{p} \lambda_k^{(j)} \lambda_k^{(\ell)}/\lambda_k^*$$

$b^{(j)}$ represent basis vectors $j = 0,1,\ldots,d$ associated with the general solution to the dual constraints (8-10)

$$\lambda_k = \sum_{i=m_k}^{n_k} \delta_i$$

for $\delta_i^* > 0$

Numerical studies testing the accuracy of the above procedures have been reported by the authors [8,9]. These studies also detail the ability of the incremental procedures to improve the accuracy of the solutions.

From the point of view of Bigelow and Shapiro [4], writing (8,9) as $A\delta = b$, changes in the coefficients satisfy the systems:

$$\dot{\pi} = \left[A \left(\frac{\partial^2 v}{\partial \delta^2} \right)^{-1} A^T \right]^{-1} \left[-A \left(\frac{\partial^2 v}{\partial \delta^2} \right)^{-1} \left(\frac{-\partial^2 v}{\partial \rho \partial \delta} \right) \dot{\rho} \right] \tag{13}$$

$$\dot{\delta} = \left(\frac{\partial^2 v}{\partial \delta^2} \right)^{-1} A^T \dot{\pi} - \left(\frac{\partial^2 v}{\partial \delta^2} \right)^{-1} \left(\frac{\partial^2 v}{\partial \rho \partial \delta} \right) \dot{\rho} \tag{14}$$

where π are the multipliers associated with $A\delta = b$, ρ is the vector of coefficients c, $\dot{\rho}$ is the vector of changing coefficients α_i^2/c_i.

Computational experience with this method has been reported in [6,9]. In both studies, an incremental procedure is evaluated since we want to be able to replace the differential changes in (11-14) with discrete approximations. The incremental procedure provides a mechanism for controlling the error resulting from this approximation [9].

The advantage of the harmonic means approach can be seen in (11,12) and (13,14). The harmonic mean approach has the property that the A matrix and hence the $b^{(j)}$ are invariant with regard to changes in the α [11]. As a result, the sensitivity analysis procedures can be applied effectively to the harmonic mean algorithm, but not as effectively to the other condensation methods [3,10]. In the other methods, since the elements a_{ij} of A are functions of the α_i, any change in the α_i induces a large number of changes throughout the model.

Regardless of which approach is used, (10,11) or (12,13), we generate approximate solutions $\delta_i' = \delta_i^* + \Delta\delta_i$, and hence new approximations $A_H(\alpha)$ without optimally resolving the harmonic program. The modification of the harmonic mean algorithm based on these procedures will replace Step 2 of the algorithm by the statement:

Modified Step 2: Using sensitivity analysis procedures, generate a feasible solution $t^{\ell+1}$ to $A_H(\alpha^{\ell+1})$, where the parameter changes are of the form $(\alpha^{\ell} - \alpha^{\ell+1})$.

Since the harmonic mean algorithm requires a feasible point (to the original problem) in order to continue, it is essential that the point $t^{\ell+1}$ generated by the modified Step 2 is feasible. According to Theorem 3A [11], if t is feasible with respect to $A_H(\alpha)$ then t is feasible with regard to A_R. Hence we need to insure the feasibility of the solutions to $A_H(\alpha)$. As described in [9], the incremental procedure can be used to maintain feasibility. For example, if a constraint is violated for a change c_i to c_i' then finer increments $(c_i' - c_i)/N$ for some appropriate value of N will insure feasibility of the solution.

Within this framework there are two computational issues addressed by this paper. While the main issue is that of iteration skipping, these subissues directly impact on the ability of such a scheme to generate useful results. These issues are:

1. What level of incremental analysis is necessary in this setting? Earlier studies [8] have shown the impact of incremental analysis in controlling the error associated with discrete changes in parameters. However, incrementing increases the amount of computational burden, which in this setting needs to be balanced against the accuracy of the approximating program $A_H(\alpha)$.
2. How often should the program be resolved *optimally*? The harmonic algorithm is generating a sequence of approximations to the original program A_R; hence we need to be concerned about how well we are representing the problem. Since the sensitivity analysis contains inherent errors, an optimal resolving of $A_H(\alpha)$ is used to eliminate the error.

Recall that the harmonic algorithm proceeds as:

Step 1: For a feasible solution $t^{\ell+\ell}$ to $A_H(\alpha^{\ell})$, compute $\alpha_i^{\ell+1} = u_i(t^{\ell})/g_k(t^{\ell})$.

Step 2: Generate a solution $t^{\ell+1}$ to $A_H(\alpha^{\ell+1})$. If the termination criterion is satisfied, stop; if not, update ℓ and go back to Step 1.

In our iteration skipping method, we generate the solution in Step 2 by considering the change from α^{ℓ} to $\alpha^{\ell+1}$ as a parameter change $|\alpha^{\ell+1} - \alpha^{\ell}|$ and using sensitivity analysis to generate a new solution $t^{\ell+1}$.

In order to evaluate the methods we constructed five different implementations with varying degrees of consideration for each of the above issues. These implementations and their rationales are:

Implementation	Method
1	Small, 1%, sensitivity analysis increments until termination. That is, the change $\lvert\alpha^{\ell+1}\rvert$ is divided into 1% increments and the sensitivity analysis applied. At the solution $t^{\ell+1}$ a new approximating program is constructed and the above analysis repeated.
2	Large, 10%, sensitivity analysis increments until termination. This is similar to 1 except we are using larger increments. *Rationale:* This method should provide a less expensive implementation than 1. However, since the increments are larger there will be more error in the approximating solutions. Also feasibility needs to be monitored and if violated, the procedure returns to the previous iteration with smaller increments. The solution is generated without solving A_H optimally.
3	Small increments, 1%, with sensitivity analysis, then resolve optimally after 10 changes. This implementation will generate optimal approximation after 10 sensitivity analysis iterations. *Rationale:* This method will eliminate the accumulated error in the approximations by the optimal resolve.
4	Large increments with sensitivity analysis then resolve optimally after 10 changes. *Rationale:* Same as 3 but with large sensitivity analysis increments.
5	Large increments, 10%, with sensitivity analysis; resolve optimally after 5 changes. *Rationale:* Same as 4, more frequent optimal resolves.

These implementations were tested on the six test problems given in the appendix. These problems were chosen because they were used in an earlier study on the harmonic mean algorithm and they represent a variety of problem types. Problems 2 and 6 each have a single reversed constraint with few terms and should not present any computational difficulties. Problem 1 has five sets of α_i to be changed at each iteration and should provide a good test of having to make many simultaneous changes. Similarly, in problem 3 there are a number of changes to be made but all within a single constraint. Problems 4 and 5 should provide a good test of the numerical stability of the procedures. Both are well known test problems which present numerical difficulties.

In each of these implementations, two additional conditions, which impact directly on the computational effort, were imposed:

1. The value of TOLERANCE, discussed earlier, was set at 1×10^{-3}. That is, termination occurred whenever $|\alpha_i^{\ell} - \alpha_i^{\ell+1}| \leqq 1 \times 10^{-3}$ for all i.
2. Each implementation terminates with a harmonic algorithm iteration. That is, based on the final values of the α_i, the optimal solution for that set of α_i is computed. It is this solution that is called the final solution.

The following tables provide a basis for comparison of these implementations in light of the preceding trade-offs. While these tables refer specifically to the above implementations, they demonstrate the feasibility of the approach in a more general context.

Table 1 details the number of iterations and amount of time for termination, and as such provides a means of comparing the amount of work involved in using the methods of Duffin, (11,12), and Bigelow and Shapiro, (13,14), as well as each the implementations of the iteration skipping scheme.

Using the results of Table 1, we can rank the various methods based on the harmonic algorithm time. These rankings are given in Table 2, and based on average results, the implementations are ranked as (from least to most expensive):

2 < 4 < 5 < Harmonic < 1 < 3

Beyond the implications of the above results, we note that the algorithm based on the Bigelow and Shapiro [4] approach, (13,14), takes advantage of special problem structure. The implications of this will be explored in detail elsewhere.

As noted previously, the results of Table 1 describe only one dimension of the problem; in addition, we need to examine the accuracy of the

Table 1 Comparison of Computational Effort

Problem Number	Harmonic Algorithm	Implementation of Iteration Skipping				
		No Resolve		Resolve After 10 Iterations*		Resolve After 5 Iterations*
		Small Increments	Large Increments	Small Increments	Large Increments	Large Increments
1	12	11	9	14 1	21 2	16 3
	13.5	42.40 28.56	4.97 4.38	54.93 37.23	11.1 8.43	9.75 7.74
2	17	17	18	18 1	19 1	18 3
	1.7	1.65 1.86	.50 .59	1.85 1.82	.62 .66	.84 .84
3	18	18	18	18 1	18 1	18 3
	2.1	1.42 7.07	.58 1.33	1.56 7.16	.62 1.42	.94 1.61
4	119	138	103	119 11	122 12	121 24
	4.0	8.57 14.47	1.81 2.08	8.35 13.04	3.18 3.02	4.45 3.12

5	19	15	13	20 2	16 1	16 3
	2.9	§	2.46 .99	23.40 9.73	3.12 1.02	3.47 2.20
6	12	11	11	10 1	10 1	11 2
	.44	.34 1.10	.19 .32	.31 1.00	.18 .27	.22 .34
Format	# of Iterations	# of Iterations		# of Iterations # of Resolves		
	CPU Time†	CPU Time Using (11,12) CPU Time Using (13,14)		CPU Time Using (11,12) CPU Time Using (13,14)		

*Iterations refers to sensitivity analysis iterations.

†All CPU time refers to seconds on an AMDAHL 470-V-6.

§Time not recorded due to unnatural termination of algorithm.

Table 2 Relative (to Harmonic Algorithm) CPU Times

Problem Number	Implementation									
	1		2		3		4		5	
	(11,12)*	(13,14)†	(11,12)*	(13,14)†	(11,12)*	(13,14)†	(11,12)*	(13,14)†	(11,12)*	(13,14)†
1	3.14	2.12	.37	.32	4.07	2.76	.82	.62	.72	.57
2	.97	1.09	.29	.35	1.09	1.07	.36	.39	.49	.49
3	.68	3.37	.28	.63	.72	3.41	.28	.68	.45	.77
4	2.13	3.62	.45	.52	2.09	3.26	.80	.76	1.11	.78
5	--	--	.85	.34	8.07	3.36	1.08	.35	1.19	.76
6	.77	2.50	.43	.77	.71	2.27	.41	.61	.50	.77
Average	1.54	2.54	.45	.48	2.79	2.69	.63	.57	.74	.69

*Using the Duffin, *et al.* approach, (11,12).

†Using the Bigelow and Shapiro approach, (13,14).

results. In all cases except one, the algorithms terminated with a solution that satisfied the above termination criterion. Since these solutions were generated using an approximation method, we want to evaluate how well they have approximated the optimal solution. The method for assessing this accuracy is to compare the true solution to the sensitivity analysis generated solution [8]. We report this as percent error, computed as:

$$\frac{\text{True Solution} - \text{Predicted Solution}}{\text{True Solution}} \times 100\%$$

These results are summarized in Table 3 and aggregated in terms of absolute deviations of the percent error in Table 4. This measure is computed for both the variables and the α_i, and while an aggregate measure, in most cases, it reflects the overall performance of the algorithms.

On the basis of the average results, the implementations would be ranked from most to least accurate as:

$$\boxed{5 > 3 > 4 > 1 > 2}$$

However, it is interesting to note the effect of problem 5 on these results. Without problem 5, methods 3 and 5 are interchanged in terms of accuracy. Problem 5, while it terminated naturally, represents a problem that is unstable with regard to the harmonic algorithm.

IV. SUMMARY AND CONCLUSIONS

The major impact of the numerical results of the previous section is to demonstrate that acceleration of the harmonic mean algorithm based on sensitivity analysis procedures is a viable alternative to the original algorithm [11]. While the effectiveness of such a scheme depends upon the particular implementation and the nature of the problem, we have demonstrated the ability of such procedures to generate meaningful results. The iteration skipping methods are shown to be less expensive computationally and can be structured to keep the error within reasonable bounds.

The computational experience of this paper indicates that based on the criteria of computational effort and deviation or error from the true result, implementation 5 is the best choice. This implementation used a strategy of large sensitivity analysis steps with more frequent resolves of the harmonic mean problem. This implementation, based on the average results of Tables 2 and 4, represents a reasonable trade-off between

Table 3 Error Analysis [% Error = (True - Predicted)/True × 100%]

Variable	Implementation				
	1	2	3	4	5
	Problem 1				
t_1	2.76	18.83	.31	.02	.18
t_2	-1.24	-3.15	.12	1.41	.53
t_3	-.004	-1.89	-.07	-.38	-.16
t_4	.75	5.61	.08	.007	.05
t_5	.001	1.20	.05	.26	.11
t_6	-.63	-4.65	-.07	-.005	-.04
t_7	.48	2.35	.004	-.27	-.08
t_8	0	.91	.04	.19	.08
t_9	.24	1.77	.03	.002	.02
t_{10}	0	.40	.02	.09	.04
t_{11}	1.31	9.24	.14	.01	.09
t_{12}	-.19	1.78	.11	.73	.29
t_{13}	-.002	-.37	-.01	-.06	-.03
$g_0(t)$	-.019	-.46	0	0	0
	Problem 2				
t_1	.004	.03	-.005	-.003	-.003
t_2	.190	1.18	-.23	-.17	-.15
g_0	-.005	-.05	.006	.005	.004
	Problem 3				
t_1	0	0	0	0	0
t_2	.06	.42	.01	.09	.03
t_3	.04	.21	.007	.06	.02
t_4	-2.24	20.53	-.43	-4.19	-1.17
t_5	-2.26	-20.79	-.43	-4.23	-1.19
t_6	.23	2.30	.04	.46	.12
t_7	.21	2.10	.04	.41	.11
g_0	-.003	-.05	-.0007	-.006	-.0007
	Problem 4				
t_1	-21.56	-24.51	.13	-.48	.13
t_2	-47.96	-40.89	.27	-.96	.27
g_0	4.67	98.02	-.05	.18	-.05

Table 3 (Continued)

Variable	Implementation				
	1	2	3	4	5
			Problem 5		
t_1	1.62	.126	-2.37	-8.87	-1.62
t_2	3.09	-16.56	-4.20	-14.87	-2.51
t_3	1.19	-13.18	-2.51	-9.08	-1.82
t_4	2.49	-18.65	-4.52	-15.39	-3.05
g_0	98.25	99.26	-244.1	47.56	-1.70
			Problem 6		
t_1	.053	.59	-.08	.46	-.12
t_2	.113	1.17	.11	1.17	0
t_3	.001	.06	-.13	-.08	-.12
t_4	-.02	-.27	.11	-.14	.12
g_0	-.12	-1.29	-.13	-1.29	0

accuracy and computational effort. One of the lessons to be learned from these results is the need to resolve the harmonic approximating problem in order to control the error being generated in the approximations. This resolving can be accomplished with a minimal impact on the computational burden, and while the question of an optimal frequency of resolving is probably unanswerable, our results here point to some usable implementations.

We note that the problems tested here involve a wide range of difficulty from relatively simple to the potential difficulties of problem 5. This problem, Rosenbrock's problem, is a difficult problem for the harmonic approach in that it is extremely sensitive to choices of the value of tolerance. The implementations of the iteration skipping algorithm had some difficulty with the problem, but with appropriate resolves of the harmonic approximation, these methods performed as well as the basic harmonic mean algorithm.

Finally, a word about the two different methods of performing the sensitivity analysis. As mentioned earlier, it is not our purpose to compare these approaches to sensitivity analysis. However, there are some observations that we can make as the result of this study. As shown by the result of problem 1 in Table 1, the Bigelow and Shapiro approach does offer

Table 4 Absolute Deviations of Percent Error

Problem Number	1		2		3		4		5	
	Variables	α's	Variables	α's	Variables	α's	Variables	α's	Variables	α's
1	7.61	5.38	52.15	39.01	1.05	.43	3.43	1.93	1.70	1.05
2	.19	.43	1.21	2.76	.24	.37	.17	.26	.15	.26
3	5.04	2.16	46.35	1.99	.96	.48	9.44	3.59	2.64	.89
4	69.52	104.73	65.40	204.69	.40	.94	1.44	3.47	.40	5.65
5	8.39	112.04	48.52	93.65	13.60	71.30	48.21	44.60	9.00	18.34
6	.18	.002	2.09	.33	.43	.65	1.85	.34	.36	.25
Average	15.16	37.46	35.95	57.07	2.78	12.36	10.76	9.03	2.38	4.41

some computational advantage when dealing with larger problems. The number of iterations and error analysis are slightly improved using the Bigelow and Shapiro approach. However, a more complete study is necessary to quantify these statements.

APPENDIX: Test Problems

1. *Source:* Avriel and Williams [3]

   Problem characteristics: Five reversed constraints and hence, five sets of α_i changing at each iteration.

$$
\begin{aligned}
\min \quad & t_1 + t_2 + t_3 \\
\text{s.t.} \quad & t_4 + t_6 \leqq 400 \\
& t_5 + t_7 \leqq 400 + t_4 \\
& 500 + t_8 \leqq 600 + t_5 \\
& t_4 + .12t_1 \leqq 100 + .0012t_1t_6 \\
& t_5 + .0008t_2t_4 \leqq t_4 + .0008t_2t_7 \\
& 500 + .0004t_3t_5 \leqq t_5 + .0004t_3t_8 \\
& t_j > 0 \qquad t_j = 1,\ldots,8
\end{aligned}
$$

Starting point: t = (5000, 5000, 5000, 200, 350, 150, 250, 425)

Optimal solution:

$t_1 = 578.963$	$t_4 = 181.988$	$t_7 = 286.382$
$t_2 = 1360.437$	$t_5 = 295.606$	$t_8 = 395.606$
$t_3 = 5109.848$	$t_6 = 218.011$	$g_0(t) = 7049.249$

2. *Source:* Dinkel, *et al.* [7]

   Problem characteristics: Approximate the objective function and one constraint; relatively smooth problem.

$$
\begin{aligned}
\min \quad & 2.5t_1^{.54}t_2^{1.46} + 10386t_2 + 80t_1 - 10000t_1^{.27}t_2^{.73} \\
\text{s.t.} \quad & 100t_1^{-.73}t_2^{.73} - .025t_1^{-.46}t_2^{1.46} - 103.865t_1^{-1}t_2 \leqq 1 \\
& t_1, t_2 > 0
\end{aligned}
$$

Starting point: t = (50000, 200)

Optimal solution: $g_0(t) = 1073032.656$

$t_1 = 53641.019$

$t_2 = 317.153$

3. *Source:* Dinkel and Kochenberger [6]

   Problem characteristics: Maximize operator results in reversed constraint.

$$\max \quad 1.1t_1^{.51} + t_2^{.47}t_3^{.24} + .9t_1^{.51}t_4^{.53}t_5^{.19} + 1.4t_1^{.51}t_6^{.5}t_7^{.21}$$

$$\text{s.t.} \quad 50t_2 + 120t_4 + 85t_6 \leqq 10000$$

$$t_1 \leqq 1000$$

$$t_3 + t_5 + t_7 \leqq 500$$

$$t > 0$$

   Starting point: $t_j = 1$, $j = 1,\ldots,7$

   Optimal solution: $g_0(t) = 2420.284$

   $t = (1000, 99.91, 276.46, 4.629, 21.557, 52.343, 202.297)$

4. *Source:* This is the Rosenbrock problem; see Himmelblau [17, Appendix A, Problem 2]

   Problem characteristics: Commonly used test problem; steep curved valley (Banana) along $x_2 = x_1^2$.

$$\min \; 100(t_1^2 - t_2)^2 + (1 - t_1)^2 + 1$$

$$t_1, t_2 > 0$$

   Starting point: $t_1 = .1$, $t_2 = .2$

   Optimal solution: $g_0(t) = 1$, $t_1 = t_2 = 1$

5. *Source:* Himmelblau [17, Appendix A, Problem 8]

   Problem characteristics: Nonoptimal stationary point at $g_0(t) = 8$.

$$\min \; 100(t_2 - t_1^2)^2 + (1 - t_1)^2 + 90(t_4 - t_3^2)^2 + (1 - t_3)^2$$
$$+ 10.1(t_2 - 1)^2 + 10.1(t_4 - 1)^2$$
$$+ 19.8(t_2 - 1)(t_4 - 1) + 1$$

   Starting point: $t = (.4, .4, .6, .7)$

   Optimal solution: $g_0(t) = 1$, $t = (1,1,1,1)$

6. *Source:* Hayes [16]

   Problem characteristics: Smooth problem; one reversed constraint.

$$\min \quad 300t_2^{-1} + t_2^{-1}t_3^2 + t_2^{-1}t_4^2$$

$$\text{s.t.} \quad t_1^{-1}t_2t_4^{-1} \leqq 1$$

$$t_1^{-1} + t_1^{-1}t_3t_4^{-1} \geqq 1$$

Starting point: $t_1 = t_4 = 3$, $t_2 = 9$, $t_3 = 6$

Optimal solution: $g_0(t) = 24.494897$

$t \quad = (2, 24.494897, 12.24744, 12.24744)$

REFERENCES

1. R. L. Armacost and A. V. Fiacco. Computational experience in sensitivity analysis for nonlinear programming, *Mathematical Programming*, Vol. 6, pp. 301-326 (1974).
2. R. L. Armacost and A. V. Fiacco. Sensitivity analysis for parametric nonlinear programming using penalty methods, Serial T-340, Program in Logistics, The George Washington University (1976).
3. M. Avriel and A. C. Williams. Complementary geometric programming, *SIAM Journal of Applied Mathematics*, Vol. 19, pp. 125-141 (1970).
4. J. H. Bigelow and N. Z. Shapiro. Implicit function theorems for mathematical programming and systems of inequalities, *Mathematical Programming*, Vol. 6, pp. 141-156 (1974).
5. J. J. Dinkel and G. A. Kochenberger. On sensitivity analysis in geometric programming, *Operations Research*, Vol. 25, pp. 155-163 (1977).
6. J. J. Dinkel and G. A. Kochenberger. Constrained entropy models: solvability and sensitivity, *Management Science*, Vol. 25, pp. 555-564 (1979).
7. J. J. Dinkel, G. A. Kochenberger, and B. McCarl. Computational study of methods for solving polynomial geometric programs, *Journal of Optimization Theory and Applications*, Vol. 19, pp. 233-259 (1976).
8. J. J. Dinkel, G. A. Kochenberger, and S. N. Wong. Sensitivity analysis in geometric programming: computational aspects, *Transactions on Mathematical Software — ACM*, Vol. 4, pp. 1-14 (1978).
9. J. J. Dinkel, G. A. Kochenberger, and S. N. Wong. Parametric analysis in geometric programming: an incremental analysis approach, in *First Symposium on Mathematical Programming with Data Perturbations* (A. V. Fiacco, ed.), Marcel Dekker, New York, 1982.
10. R. J. Duffin. Linearizing geometric programs, *SIAM Review*, Vol. 12, pp. 211-227 (1970).
11. R. J. Duffin and E. L. Peterson. Reversed geometric programs treated by harmonic means, *Indiana University Mathematics Journal*, Vol. 22, pp. 531-550 (1972).
12. R. J. Duffin and E. L. Peterson. Geometric programming with signomials, *Journal of Optimization Theory and Applications*, Vol. 11, pp. 3-35 (1973).

13. R. J. Duffin, E. L. Peterson, and C. Zener. *Geometric Programming*. Wiley, New York, 1967.

14. A. V. Fiacco. Sensitivity analysis for nonlinear programming using penalty methods, *Mathematical Programming*, Vol. 10, pp. 287-311 (1976).

15. A. V. Fiacco and G. P. McCormick. *Nonlinear Programming: Sequential Unconstrained Minimization Technique*. Wiley, New York, 1968.

16. P. L. Hayes. *Mathematical Methods in the Social and Managerial Sciences*. Wiley Interscience, New York, 1975.

17. D. M. Himmelblau. *Applied Nonlinear Programming*. McGraw-Hill, New York, 1973.

CHAPTER 8 LEAST SQUARES OPTIMIZATION WITH IMPLICIT MODEL EQUATIONS

AIVARS CELMIŅŠ / Ballistic Research Laboratory, Aberdeen Proving Ground, Maryland

ABSTRACT

This article is concerned with algorithmic aspects of nonlinear least squares model fitting problems. Such problems differ from general optimization problems due to the special structure of the least squares normal equations. In simple least squares problems, that structure allows one to partition the normal equation system, whereby significant algorithmic simplification can be achieved. The more general least squares problems, considered here, often can be made partitionable by proper parameter manipulations. In this article, sufficient partitionability conditions are derived and parameter manipulations are discussed for the achievement of partitionability.

I. INTRODUCTION

We consider least squares model fitting tasks that can be formulated as constrained minimization problems. The constraints representing the model equations are, in general, implicit nonlinear relations between observables and parameters. The solution of such problems can be found by the Lagrange multiplier technique which provides a system of coupled nonlinear normal equations. The equations determine the optimal values of the model parameters and residuals, and can be used to analyze the sensitivity of the solution to data perturbations. However, the numerical solution of the system is not necessarily trivial, because the size of the system is proportional to the often very large number of data. Fortunately, many typical least squares model fitting problems have such a normal equation structure that the equation system can be partitioned by manipulations of the model equations. Manipulations of parameters, e.g., an introduction of new parameters and/or a formal elimination of some parameters, can be particularly effective. We will consider the effects of such manipulations and derive partitionability conditions that can be used as a guide for a rational model equation formulation, and for a rational planning of experiments.

In Section II we will give a formal definition of the least squares model fitting problem, and establish the normal equations. Algorithms for the numerical solution of the equations will be outlined in Section III, where also the sensitivity of the solution to data perturbations is investigated. Partitionability of the normal equations and corresponding parameter manipulation are discussed in Sections IV and V, respectively. Section VI gives an example for partitioning of normal equations arising in the adjustment of a planimetric traverse.

II. THE MODEL FITTING PROBLEM

Let the mathematical model of an observable event be formulated by a set of r independent equations, say,

$$F(x,t) = 0 \tag{1}$$

where $F \in R^r$ is a vector function, $x \in R^n$ represents potential observations, and $t \in R^p$ is a vector of free parameters. Equation (1) may be prescribed by a theory of the event, or chosen by other considerations. A model fitting problem arises when one seeks to determine the validity of the mathematical description by testing equation (1) with experimental data. In such tests, typical magnitudes of the dimensions p, r, and n are 10, 100, and 1000, respectively. We assume that the dimensions satisfy the inequalities

$$0 \leq p < r \leq n \tag{2}$$

which assure that the optimization problem has sufficient degrees of freedom.

We restrict our considerations to model functions F that are twice differentiable with respect to all its n + p arguments. Differentiability of model functions is typical for many problems in engineering, physics, geodesy, and other fields of application. For the present analysis, it has to be assumed only within a neighborhood of the adjusted observations x and of the optimal parameter t.

Let $X \in R^n$ be the vector of the actual observations. Because X contains observational inaccuracies one cannot expect that the theoretical description of the event, i.e., equation (1), is satisfied at x = X. Instead, one adds corrections (residuals) c to the observations and replaces the model equation (1) by the constraint equations

$$F(X + c, t) = 0 \tag{3}$$

Equation (3) means that we expect the theoretical relations involving observables and parameters to be satisfied at a vicinity X + c of the actual observations X.

Obviously, one would like the residuals c to be as small as possible. Hence, a general model fitting problem can be formulated as the following constrained minimization task

$$\begin{aligned} &\text{Minimize} \quad \|c\| \\ &\text{subject to } F(X + c, t) = 0 \end{aligned} \tag{4}$$

The solution of this problem, of course, depends on the definition of the norm $\|c\|$. In this article we consider only elliptic vector norms, defined by

$$\|c\| = \sqrt{c^T M c} \tag{5}$$

where M is a positive definite matrix. One reasonable choice of M is a diagonal matrix where the elements are proportional to the inverse squares of estimated standard errors of the observed components [3,11,12,14,25]. With this choice, the norm $\|c\|$ becomes dimensionless and the corresponding minimization problem is called "weighted least squares." A more general choice for M is the inverse of an estimated variance-covariance matrix R of the observations X [1,4,5,6,8,15,24]. If the observational errors of X are normally distributed, and the model is linear, then the use of this norm, i.e., of

$$\|c\| = \sqrt{c^T R^{-1} c} \tag{6}$$

in equation (4) produces a maximum likelihood result [7,8,15]. The norm

(6) is, of course, also dimensionless, and it includes the weighted least squares norm as a special case.

Using the norm (6) we formulate a general least squares model fitting problem as follows:

$$\begin{aligned} &\text{Minimize} \quad W = ||c||^2 = c^T R^{-1} c \\ &\text{subject to } F(X + c, t) = 0 \end{aligned} \tag{7}$$

The unknowns of the problem are the residual vector c and the parameter vector t. Given are the vector of observations X, the model function F, and an estimated variance-covariance matrix R. The latter needs to be known only up to a factor, because the inclusion of an arbitrary factor in W does not change the minimum location. Also, the model function F can be manipulated as long as the result produces mathematically equivalent constraints.

Problem (7) can be reduced to an unconstrained minimization problem if the residuals c can be eliminated from W by using the constraints. This is the case, for example, when the constraints are explicit in terms of X + c. Many least squares algorithms have been devised to treat this special problem [1-5,11,17,21-23]. In this article, we consider the more general situation where F is a nonlinear function of c and t such that a formal elimination of all or some residuals is either not possible or not practical.

In order to simplify our notation in the subsequent analysis, we shall denote derivatives of F by subscripts. Thus, e.g.,

$$\frac{\partial F(X + c, t)}{\partial X} = F_X(X + c, t)$$

is an $r \times n$ matrix, and

$$\frac{\partial^2 (K^T F(X + c, t))}{\partial X \partial t} = (K^T F(X + c, t))_{Xt}$$

where $K \in R^r$, is an $n \times p$ matrix.

In addition to the differentiability of F we also assume that in a neighborhood of the solution (X + c,t) the following rank conditions are satisfied:

$$\text{rank } F_X = r \tag{8}$$

and

$$\text{rank } F_t = p \tag{9}$$

The condition (8) insures that the model equations (3) are independent, and the condition (9) excludes model formulations with redundant parameters.

Next, we obtain normal equations (necessary conditions) for the problem (7) using Lagrange multipliers. Let $k \varepsilon R^r$ be a vector of correlates (Lagrange multipliers), and let

$$\tilde{W} = 1/2\ c^T R^{-1} c - k^T F(X + c,t) \tag{10}$$

be the modified objective function. By setting the derivatives of $\tilde{W}$ with respect to c, t, and k equal to zero, we obtain the equations [7,9,16,20]

$$c - R \cdot F_X^T(X + c,t) \cdot k = 0 \tag{11a}$$

$$k^T \cdot F_t(X + c,t) = 0 \tag{11b}$$

$$F(X + c,t) = 0 \tag{11c}$$

The normal equations (11) generally have several solutions, one of which is the solution of the optimization problem (7). The selection of the proper solution is normally done by an analysis of problem related background information. We shall not discuss such analyses in this article, and concentrate instead on the finding of any numerical solution of equations (11).

III. ITERATION ALGORITHMS AND EFFECTS OF DATA PERTURBATION

The normal equations (11) are nonlinear with respect to the unknowns c and t. Therefore, their numerical solution will generally require an iteration. One obtains second order Newton-type iteration equations by expanding the normal equations at an approximate solution.

Let C, K, and T be approximations to the solution vectors c, k, and t, respectively, and ε, κ, and τ be the corresponding corrections. The linear terms of an expansion of equation (11) at the approximate solution produce the following system of Newton-Raphson equations for the corrections [20]:

$$\left.\begin{aligned}
&[I - R \cdot (K^T \cdot F)_{XX}] \cdot \varepsilon - R \cdot F_X^T \cdot (K + \kappa) - R \cdot (K^T \cdot F)_{Xt} \cdot \tau = -C \\
&(K^T \cdot F)_{tX} \cdot \varepsilon + F_t^T\ (K + \kappa) + (K^T \cdot F)_{tt} \cdot \tau = 0 \\
&F_X \cdot \varepsilon + F_t \cdot \tau = -F
\end{aligned}\right\} \tag{12}$$

The arguments of F and of its derivatives in equation (12) are the approximations X + C and T.

An iteration based on equation (12) proceeds by computing the corrections ε, κ, and τ, adding them to the approximations C, K, and T, respectively, and repeating the process. The equations may be rearranged into a more convenient form for the iteration. An example of such rearranged iteration equations is given in the appendix and corresponding computer programs are described in [10].

An often used variation of the Newton-Raphson equations is obtained by setting in equation (12) all second order derivatives of F equal to zero [2-4,6-8,12,13,15,16,21-23,25]. The resulting equations are called Gauss-Newton equations. Iteration algorithms based on Gauss-Newton equations converge only linearly and may have other disadvantages [20].

The linear terms of the expansion of the normal equations, i.e., the equation system (12), also provide a means to obtain estimates of the effects of data perturbations on the solution. To show this, we express the normal equations in terms of the corrected observations x = X + C, obtaining the system

$$\left.\begin{aligned} & x - R\cdot F_x^T(x,t)\cdot k = X \\ & k^T\cdot F_t(x,t) = 0 \\ & F(x,t) = 0 \end{aligned}\right\} \qquad (13)$$

which we expand at the solution. The linear terms of the expansion yield the following relation between the differentials of the solution x, k, t, and the differentials of the observations X:

$$\left.\begin{aligned} & [I - R\cdot(k^T\cdot F)_{xx}]\,dx - R\cdot F_x^T df - R\cdot(k^T\cdot F)_{xt}dt = dX \\ & (k^T\cdot F)_{tx}dx + F_t^T dk + (k^T\cdot F)_{tt}dt = 0 \\ & F_x dx + F_t dt = 0 \end{aligned}\right\} \qquad (14)$$

The coefficient matrices in equation (14) are identical to those in equation (12), except that now the functions are evaluated at the solution. Therefore, changes of the solution corresponding to small data perturbations dX can be calculated, using, for example, the same programs as for the iteration equations of the appendix. Thus, one obtains the changes dt of the parameters corresponding to the perturbations dX from the equation

$$N\,dt = S\,dX \qquad (15)$$

where N and S are defined in the appendix in terms of F and its first and second order derivatives. A formula for corresponding changes dx of the adjusted observations is

$$dx = \Gamma\ A\ dX - E_1 dt \tag{16}$$

where Γ, A, and E_1 are defined in the appendix.

The significance of the perturbations equations (15) and (16) is that they establish explicit algorithms for the computation of data perturbation effects in problems with implicit nonlinear models. Traditionally, model sensitivity is investigated by assuming that the model is linear or linearizable [25]. However, as we have shown, the first order effects of perturbations do depend on second order derivatives of the model. That dependency is lost if the model has been linearized at the outset. The numerical significance of the second order derivative terms depends on the model function F as well as on the data and, therefore, cannot be estimated in general.

It is obvious that the algorithms for the computation of the perturbations dt and dx can be simplified by the same manipulations of the model equations that make the normal equations partitionable. We discuss such manipulations in the next section.

In most model fitting problems one is interested to obtain an estimate of the accuracy of the fitted model in terms of estimated data accuracies. In our formulation, the data accuracy is represented by the data variance-covariance matrix R. The perturbation equation (15) permits one to derive a formula of a corresponding estimate of the variance-covariance matrix V_t of the parameters. The formula is obtained by applying the law of variance propagation to (15) with the result [9,10,20]

$$V_t = N^{-1} SRS^T (N^{-1})^T \tag{17}$$

In case the variance-covariance matrix R of the data has been estimated only up to a factor, the formula must be supplemented by an estimate of that factor. The usual estimate is

$$m_0^2 = \frac{1}{n-p}\, c^T R^{-1} c \tag{18}$$

The square root m_0 of the factor is also called the standard error of weight one.

It is clear from the derivation of (17) that the formula for the variances of t contains first and second order derivatives of the model function F in spite of the fact that the formula is only a first order estimate of the variances. (The dependence on the second order derivatives is shown explicitly in the appendix.) Authors who prefer Gauss-Newton algorithms for the numerical solution of the normal equations tend to overlook this fact and present variance estimate formulas without second order derivatives of the model function F [2-4,6-8,12,13,15,16,19,21,23-25]. Such formulas are less than first order accurate and should not be used without an estimate of the effect of the neglected terms. One can easily construct examples in which the second order derivative terms contribute significantly to V_t, either increasing or decreasing the estimated variances. In some applications, the inadequacy of the simplified variance estimate formulas is common knowledge [18], but few authors warn their readers about the limited applicability of their simplified formulas [6,7,13,15,24]; however, they do not derive the complete expressions (17).

The computation of V_t by the formula (17) again can be done with very little effort if the normal equations are solved by a Newton-Raphson algorithm. The parameter accuracy estimates then can be calculated concurrently with the parameters, and may be used in the iteration end criterion [10]. If a Gauss-Newton algorithm is used, then one needs a separate program that calculates the variance-covariance matrix, e.g., after a solution of the normal equation has been found. In any case, the computations are simplified significantly if the problem can be partitioned.

The variance-covariance matrix V_t is the key to any further investigation of the accuracy and sensitivity of the model. It is important to remember that for that purpose the off-diagonal elements of V_t are as essential as the diagonal elements, which merely provide estimates for the marginal variances of the parameters [15]. The example in Section VI illustrates the use of the matrix and the effect of the off-diagonal terms on the confidence limits of the optimization result.

IV. PARTITIONABILITY

If the data volume is large, then the numerical solution of the normal equations (11) can be a formidable task. In a typical model fitting problem the dimension n of the observations is of the order 1000 or larger, and one has to manipulate matrices of the order $n \times n$ in the Newton-Raphson iteration equations. Fortunately, many least squares problems have such a structure that the large systems of equations can be partitioned into a set of smaller systems, whereby the manipulation of the large matrices can be avoided.

We illustrate this partitionability with a curve fitting problem in the y,z-plane. Let the curve be defined by the implicit equation

$$f(x;t) = \hat{f}(y,z;t) = 0 \tag{19}$$

where x is the coordinate vector in the y,z-plane and t is a parameter vector. Let the observations X_i, i = 1,2,...,s, be the coordinates of s points in the y,z-plane, i.e.,

$$X_i = \begin{pmatrix} y_i \\ z_i \end{pmatrix} \qquad i = 1,2,\ldots,s \tag{20}$$

and let the accuracies of the observed points be characterized by corresponding estimated variance-covariance matrices

$$R_i = \begin{pmatrix} v_{iyy} & v_{iyz} \\ v_{iyz} & v_{izz} \end{pmatrix} \qquad i = 1,2,\ldots,s \tag{21}$$

Correlations between any two observed points we assume to be negligible. The correspondences between this problem and the general model fitting problem are as follows:

$$X = \begin{pmatrix} X_1 \\ \vdots \\ X_s \end{pmatrix} \tag{22}$$

$$R = \begin{pmatrix} R_1 & & 0 \\ & \ddots & \\ 0 & & R_s \end{pmatrix} \tag{23}$$

and

$$F(X,t) = \begin{pmatrix} f(X_1,t) \\ \vdots \\ f(X_s,t) \end{pmatrix} \tag{24}$$

Using these correspondences, the curve or model fitting problem (7) can be defined in terms of the subsets X_i of X, the submatrices R_i of R, and the components f of F as follows:

$$\begin{aligned} &W = \sum_{i=1}^{s} c_i^T R_i^{-1} c_i = \min \\ &\text{subject to } f(X_i + c_i,t) = 0 \qquad i = 1,2,\ldots,s \end{aligned} \tag{25}$$

The normal equations of the optimization problem (25) are

$$c_i - R_i \cdot f_X^T (X_i + c_i, t) \cdot k_i = 0 \qquad i = 1,2,\ldots,s \tag{26a}$$

$$\sum_{i=1}^{s} k_i \cdot f_t (X_i + c_i, t) = 0 \tag{26b}$$

$$f(X_i + c_i, t) = 0 \qquad i = 1,2,\ldots,s \tag{26c}$$

where k_i $(i = 1,2,\ldots,s)$ are the scalar correlates of the problem.

A comparison of equations (26) with the general normal equations (11) shows that in the curve fitting problem the large equation system is partitioned into a set of smaller systems, so that the largest dimension of matrices to be manipulated is the maximum of 2×2 and $p \times p$. Particularly, equation (26a) is a set of s systems of two equations, each system depending only on two residuals, whereas (11a) is one system of 2s equations depending on 2s residuals. Likewise, (26c) are s scalar equations, each depending on two distinct residuals, whereas the corresponding equation (11c) is a system of s coupled equations for all 2s residuals. Obviously, the numerical treatment of equations (26) is much simpler than that of equations (11).

A basic property of the sample problem (25) is that the r constraints are scalar equations (hence $r_i = 1$ and $r = \sum r_i = s$), each depending on a distinct subset X_i of the observation vector X, and that the s subsets X_i are not correlated. We call such a problem a *standard least squares problem* because of its common occurrence and simplicity. (Most least squares literature is restricted to standard problems or to subsets of standard problems, e.g., [1-5,11,13-15,17-19,21-23].) Standard least squares problems are easier to solve numerically than general problems, because the maximum dimensions of matrices in the normal equations are independent of the total number of observations. A problem with the latter property we call *totally partitionable*. Hence, a standard least squares problem is totally partitionable. If the data are not correlated, then any fitting of a hypersurface to points in a space of observables is a totally partitionable problem. Such a fitting in, say, an m-dimensional space is also a standard problem if the dimension of the hypersurface is $m - 1$.

Next, we derive conditions for partitionability of the normal equations of general problems by comparing the structures of equations (11) and (26). First, we notice that in order to be able to partition (11c) at all, the model function F must be transformable into such a form that subsets of components of F depend on distinct subsets of the observations X. This property can be conveniently expressed by the requirement that the Jacobian matrix $\partial F/\partial X$ has a stretched block diagonal structure, i.e.,

$$\frac{\partial F}{\partial X} = \begin{pmatrix} \partial F_1/\partial X_1 & & 0 \\ & \ddots & \\ 0 & & \partial F_s/\partial X_s \end{pmatrix} \tag{27}$$

In the sample curve fitting problem (25), F has this property, whereby the submatrices $\partial F_i/\partial X_i$ are the two-component vectors $\partial f/\partial X_i$. In more general situations the submatrices have dimensions $r_i \times n_i$ and the r_i and n_i can be different for different indexes i. Because $\partial F/\partial X$ is an $r \times n$ matrix, then obviously $\sum r_i = r$ and $\sum n_i = n$.

The stretched block diagonal structure (27) of $\partial F/\partial X$ suffices to partition equation (11c). In order to partition (11a) too, one needs an additional condition on the variance-covariance matrix R. If R is diagonal and (27) holds, the equation (11a) is partitionable. However, for partitionability it is already sufficient that R have a block diagonal structure

$$R = \begin{pmatrix} R_1 & & 0 \\ & \ddots & \\ 0 & & R_s \end{pmatrix} \tag{28}$$

where the dimensions n_i of the submatrices R_i match the dimensions n_i of the submatrices $\partial F_i/\partial X_i$. Both of these conditions together are sufficient to partition the problem into s parts. Thus, if R and $\partial F/\partial X$ have the indicated structures, equation (11a) has the form

$$\begin{pmatrix} c_1 \\ \vdots \\ c_s \end{pmatrix} - \begin{pmatrix} R_1\left(\partial F_1/\partial X_1\right)^T & & 0 \\ & \ddots & \\ 0 & & R_s\left(\partial F_s/\partial X_s\right)^T \end{pmatrix} \begin{pmatrix} k_1 \\ \vdots \\ k_s \end{pmatrix} = 0 \tag{29}$$

where the c_i are distinct subsets of c with the dimensions n_i, and the k_i are correlate vectors with the dimensions r_i.

In summary, sufficient for the partitionability of a least squares model fitting problem is that the following two conditions hold:

a. R has a block diagonal structure (28), and
b. $\partial F/\partial X$ has a matching stretched block diagonal structure (27).

In data reduction problems one has no control over the structure of R, except during the planning stage of an experiment. Once the measurements are made, R is part of the given data basis. However, in many practical problems R is diagonal or nearly diagonal with few nonzero off-diagonal elements. In these cases, the partitionability of the problem depends on the formulation of the model equations F = 0. By a proper manipulation of the constraints one can often partition a problem that was not partitionable in the original formulation. We shall give an example of such a manipulation in Section VI.

The algorithmic advantages of partitioning cannot be overemphasized. In fact, partitionability rather than the special form of the objective function W is the practically important difference between a least squares problem and a general optimization problem. Most published algorithms for the solution of least squares problems are restricted to partitionable cases.

V. MANIPULATION OF PARAMETERS

This section gives an overview of parameter manipulations that can be used to achieve partitionability of least squares model fitting problems. First, we express the constraint equations (3) in the more convenient form

$$\bar{F}(c,t) = 0 \tag{30}$$

by including the observed X in the definition of the model function $\bar{F}$. Like equation (3), equation (30) is a set of r scalar equations. In general, each of the r equations depends on different subsets of the components of c and t. A subset t_e of the parameter vector t we define as the vector of *essential parameters* if all components of t_e appear in all r equations (30). All other parameters we call *nonessential*. A constraint formulation that contains only essential parameters we call a *minimal formulation*.

Minimal formulations of constraints are important in the context of partitionability. To illustrate this, we notice that the numbers and types of parameters are not intrinsic properties of a model fitting problem. Parameters can be eliminated and added, within limits, without changing the solution of the problem. However, from an algorithmic viewpoint it is not advisable to eliminate essential parameters, i.e., to reduce the problem formulation below a minimal formulation.

We illustrate this remark by considering a standard problem with a diagonal variance-covariance matrix R and only essential parameters. Let the constraints (30) of the problem be, componentwise,

$$\bar{f}_i(c_i,t) = 0 \qquad i = 1,2,\ldots,r \tag{31}$$

In order to eliminate the parameters we may use the last p equations (31) and express t in terms of the residual subset $c_e = (c_{r-p+1},\ldots,c_r)$. Substituting this expression into the first r - p equations, one obtains a system of constraints equivalent to the original system, but without parameters, namely,

$$\tilde{F}(c) = 0 \tag{32}$$

with the components

$$\tilde{f}_i(c_i,c_e) = 0 \qquad i = 1,\ldots,r - p \tag{33}$$

Now, the arguments of the components $\tilde{f}_i$ of $\tilde{F}$ are not distinct subsets of c and, therefore, the Jacobian matrix $\partial\tilde{F}/\partial c$ has not the necessary stretched block diagonal form. The problem is not partitionable in this formulation.

While a problem formulation with fewer constraints than in a minimal formulation certainly is not optimal for a numerical treatment, one may find, in some cases, that larger than minimal formulations are more practical. One obtains such formulations by introducing new parameters into the problem. The total number of parameters that can be added to a least squares problem is, however, limited by the inequalities (2). Let us assume that for a given problem the inequalities are satisfied, and let us introduce $\hat{p}$ new parameters into the problem. The corresponding $\hat{p}$ new equations which define the parameters are added to the set of constraint equations. Therefore, the inequality (2) for the new problem formulation is

$$0 \leqq p + \hat{p} < r + \hat{p} \leqq n \tag{34}$$

Hence, the number $\hat{p}$ of new parameters that can be introduced into a problem is limited by the condition

$$\hat{p} \leqq n - r \tag{35}$$

A "natural," application oriented formulation of a model fitting problem is not necessarily the most advantageous one for numerical treatment, and may even be subminimal, as in the following example. Suppose that some effect y of an explosion has been observed at different stations as a function of time t. Then the data basis consists of an observation T_0 of the time t_0 of the explosion and of a series of pairs (T,Y), providing the observed effects Y at times T. Let the theoretical model equation of the observed effect at station j be

$$y = f_j(t - t_0;\ \alpha,\ \beta,\ \gamma) \tag{36}$$

where α, β, and γ are essential model parameters. Then the corresponding constraint equations are

$$Y_i + c_{Yi} - f_i\left(T_i + c_{Ti} - (T_0 + c_{T0});\ \alpha,\ \beta,\ \gamma\right) = 0 \qquad i = 1,2,\ldots,r \tag{37}$$

The formulation (37) is minimal, because all parameters are essential. However, the problem is not partitionable even when all observations are uncorrelated, because all constraint equations contain the same observation T_0.

A partitioning of this problem can be easily achieved by introducing the starting time of the explosion as a fourth parameter δ. The correspondingly modified set of constraint equations (37) is

$$\left.\begin{aligned} & Y_i + c_{Yi} - f(T_i + c_{Ti} - \delta;\ \alpha,\ \beta,\ \gamma) = 0 \qquad i = 1,\ldots,r \\ & T_0 + c_{T0} - \delta = 0 \end{aligned}\right\} \tag{38}$$

Now we have r + 1 constraints, four parameters, and (m + 1)·r + 1 observations, where m = dim Y. The inequalities (34) are in this case

$$0 \leqq 4 < r + 1 \leqq (m + 1)\cdot r + 1 \tag{39}$$

indicating that one needs at least four observation sets (T,Y) to have a regular adjustment problem.

The problem formulation (38) is not minimal, because the parameters α, β, and γ are not essential. However, because they appear in all but one of the constraint equations, their elimination would not simplify the problem. If the data are not correlated, then in the formulation (38) the problem is totally partitionable. It is, in fact, a standard problem, if the observed effects Y_i are scalar, i.e., if m = 1.

The goal of manipulation of the model equations is to obtain an equation system with a stretched block diagonal Jacobian matrix. If the model equations are linear, then this can be achieved by algebraic manipulations. For problems with nonlinear implicit model equations, probably the most effective approach is through parameter manipulation, such as shown in the previous example. A numerical example of another problem will be given in the next section.

VI. EXAMPLE

We present as a numerical example a least squares model fitting problem arising from the adjustment of a planimetric net. The measurements in such a net have to satisfy net closure conditions, i.e., constraints without any parameters. The constraints form a set of simultaneous nonlinear equations involving all corrections, and net adjustment problems are not partitionable in this formulation. However, by introducing station coordinates as parameters, net adjustment problems always can be partitioned.

A simple specific example is the planimetric traverse shown in Figure 1. Let the observations be the distances r_i between the stations and the corresponding azimuths ϕ_i. The constraints for the closed polygon of Figure 1 are obtained from the model equations (closure conditions)

$$\left.\begin{aligned} &\sum_{i=1}^{5} r_i \sin\phi_i = 0 \\ \text{and} \quad &\\ &\sum_{i=1}^{5} r_i \cos\phi_i = 0 \end{aligned}\right\} \qquad (40)$$

by substituting in them the corrected $r_i + c_{ri}$ and $\phi_i + c_{\phi i}$ for the observed r_i and ϕ_i, respectively. Hence, the problem has two nonlinear scalar model equations (r = 2), ten observations (n = 10), and no parameters (p = 0). For simplicity, we assume that the observations are not correlated, i.e., that the estimated variance-covariance matrix R is diagonal. Nevertheless, the adjustment problem is not partitionable in this formulation.

Next, we introduce, as parameters, the coordinates of the stations 1, 2, 3, and 4 relative to the reference station A. With these eight parameters, the two original model equations (40) can be expressed equivalently by the following five sets of two equations each:

$$\left.\begin{aligned} f_1 &= \begin{pmatrix} r_1 \sin\phi_1 - x_1 \\ r_1 \cos\phi_1 - y_1 \end{pmatrix} = \begin{pmatrix} 0 \\ 0 \end{pmatrix} & f_4 &= \begin{pmatrix} x_3 + r_4 \sin\phi_4 - x_4 \\ y_3 + r_4 \cos\phi_4 - y_4 \end{pmatrix} = \begin{pmatrix} 0 \\ 0 \end{pmatrix} \\ f_2 &= \begin{pmatrix} x_1 + r_2 \sin\phi_2 - x_2 \\ y_1 + r_2 \cos\phi_2 - y_2 \end{pmatrix} = \begin{pmatrix} 0 \\ 0 \end{pmatrix} & f_5 &= \begin{pmatrix} x_4 + r_5 \sin\phi_5 \\ y_4 + r_5 \cos\phi_5 \end{pmatrix} = \begin{pmatrix} 0 \\ 0 \end{pmatrix} \\ f_3 &= \begin{pmatrix} x_2 + r_3 \sin\phi_3 - x_3 \\ y_2 + r_3 \cos\phi_3 - y_3 \end{pmatrix} = \begin{pmatrix} 0 \\ 0 \end{pmatrix} \end{aligned}\right\} \qquad (41)$$

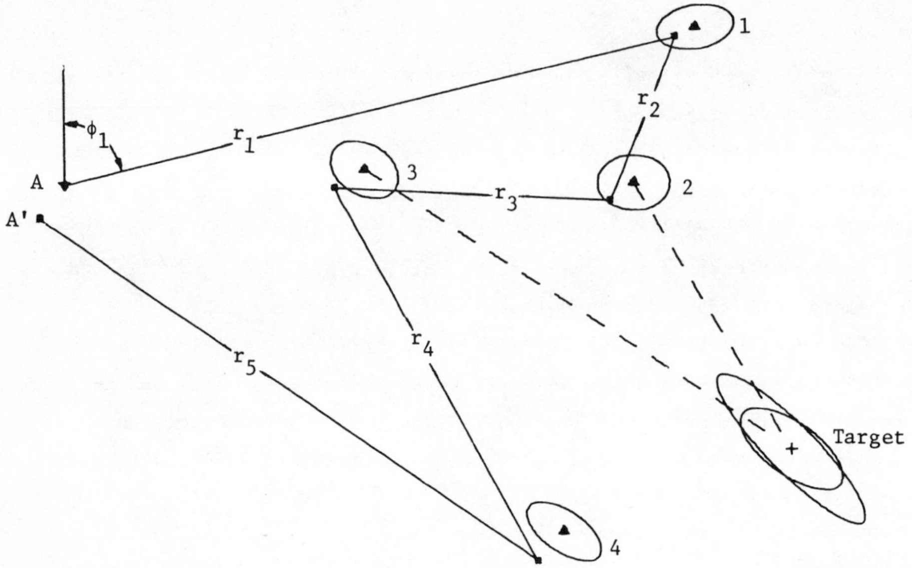

Figure 1. Planimetric Traverse

∇ - Reference station

■ - Original survey station

Δ - Adjusted survey station

\+ - Target, determined by azimuth measurements from Stations 2 and 3

The difference between Stations A and A' is the closure error of the traverse. A and A' coincide after adjustment. Accuracies of the adjusted stations and of the target are indicated by two standard error ellipses. The larger error ellipse around the target is obtained if correlations between station coordinates are neglected (Case (c) in Table 4). Data are given in Tables 1 through 4.

The corresponding constraint equations are obtained by substituting in equation (41) the corrected distances and angles for the observed ones. Now, we have $s = 5$, $r = 10$, $n = 10$, and $p = 8$. According to equation (35) no additional parameters can be introduced. In this formulation, the Jacobian matrix of the model function has the block diagonal form

$$\frac{\partial F}{\partial X} = \begin{pmatrix} \partial f_1/\partial(r_1,\phi_1) & & 0 \\ & \ddots & \\ 0 & & \partial f_5/\partial(r_5,\phi_5) \end{pmatrix} \tag{42}$$

with 2×2 matrices in the diagonal. Hence, the problem partitions into five subsets.

We note in passing that in this example the reformulation of the model equations has only slightly reduced the size of the problem. In the original formulation we would have to deal with 10×10 matrices in the normal equations, corresponding to the ten observations. The partitioning has reduced this part of the formulation to five 2×2 matrices, which can be handled much more easily. However, we have also introduced eight parameters and corresponding 8×8 matrices in the normal equations. Hence, the reduction of matrix sizes is only from 10×10 to 8×8. If numerical efficiency were important in this example, then one would introduce fewer than eight parameters, e.g., only the coordinates of stations 1 and 3. In that formulation, the largest matrix to be handled would be only 4×4.

Numerical values of the observations are given in Table 1, together with the results of the adjustments which were calculated with the computer program COLSMU of [10]. The parameter values, i.e., the station coordinates are listed in Table 2, and Table 3 provides the correlation coefficients between the station coordinates. The correlation coefficient matrix C is defined in terms of the variance-covariance matrix V_t by

$$C = D^{-1/2} V_t D^{-1/2} \tag{43}$$

where

$$D = \text{diag } V_t \tag{44}$$

Both matrices, C and V_t, were calculated by the cited utility routine which uses equation (17) for the calculation of V_t.

The adjusted positions of the stations are shown in Figure 1 together with the corresponding two standard error ellipses indicating the accuracies of the positions.

Estimates of variances and covariances of the adjusted survey stations are important when the stations are used as bases for other measurements.

Table 1 Observations and Adjustments

Nr	r(km)	e_r	c_r	$r + c_r$	$\phi(°)$	e_ϕ	c_ϕ	$\phi + c_\phi$
1	10.5	0.47	0.356	10.856	77.0	1.0	0.33	76.67
2	2.9	0.25	-0.148	2.752	202.0	1.0	-0.01	201.99
3	4.6	0.32	-0.107	4.493	273.0	1.0	0.17	273.17
4	7.0	0.40	-0.199	6.801	151.0	1.0	-0.24	150.76
5	10.1	0.46	0.045	10.145	304.0	1.0	0.42	304.42

Table 2 Adjusted Coordinates of Stations with Estimated Standard Errors

Nr	x(km)	e_x(km)	y(km)	e_y(km)
1	10.563	0.324	2.503	0.183
2	9.533	0.306	-0.049	0.228
3	5.047	0.286	0.199	0.237
4	8.369	0.310	-5.735	0.242

Weighted sum of correction squares W = 1.665 635 46
Standard error with weight one m_0 *= 0.912 589*
The factor m_0 *is not included in the standard errors of station coordinates.*

As an example, let us assume that Stations 2 and 3 are used to determine the position of a target by azimuth measurements. Let ψ_2 and ψ_3 be the azimuths observed from Stations 2 and 3, respectively. Then the target position is given by

$$x_t = x_2 - \frac{\sin\psi_2}{\sin(\psi_3 - \psi_2)}\left[(x_3 - x_2)\cos\psi_3 + (y_3 - y_2)\sin\psi_3\right]$$
$$y_t = y_2 - \frac{\cos\psi_2}{\sin(\psi_3 - \psi_2)}\left[(x_3 - x_2)\cos\psi_3 + (y_3 - y_2)\sin\psi_3\right] \quad (45)$$

Estimated variances and covariances of the target coordinates can be computed from the estimated accuracies of the azimuth observations and the variances and covariances of the bases' coordinates by applying the lin-

Table 3 Correlation Matrix of Adjusted Station Coordinates

	x_1	y_1	x_2	y_2	x_4	y_3	x_4	y_4
x_1	1.0000	0.2080	0.9572	-0.1477	0.5516	-0.1303	0.4494	-0.1801
y_1	0.2080	1.0000	0.1332	0.5161	-0.0594	0.4872	0.0394	0.1454
x_2	0.9572	0.1332	1.0000	0.0053	0.5358	0.0141	0.4626	-0.1375
y_2	-0.1477	0.5161	0.0053	1.0000	-0.2509	0.9417	-0.0407	0.3170
x_3	0.5516	-0.0594	0.5358	-0.2509	1.0000	-0.3059	0.8018	-0.3450
y_3	-0.1303	0.4872	0.0141	0.9417	-0.3059	1.0000	-0.0771	0.3470
x_4	0.4494	0.0394	0.4626	-0.0407	0.8018	-0.0771	1.0000	-0.6092
y_4	-0.1801	0.1454	-0.1375	0.3170	-0.3450	0.3470	-0.6092	1.0000

Table 4 Target Position

Observed Azimuths of Target
$\psi_2 = 149.0 \pm 0.3$
$\psi_3 = 123.0 \pm 0.3$
Computed Coordinates of Target
a. All covariances considered
$x_t = 12.159 \pm 0.437$
$y_t = -4.419 \pm 0.337$
Correlation coefficient $c_{xy} = -0.667256$
b. Covariances between stations neglected
$x_t = 12.159 \pm 0.609$
$y_t = -4.419 \pm 0.555$
Correlation coefficient $c_{xy} = -0.906523$
c. All covariances neglected
$x_t = 12.159 \pm 0.629$
$y_t = -4.419 \pm 0.616$
Correlation coefficient $c_{xy} = -0.900271$

earized law of variance propagation to equation (45). The results are shown in Table 4 and Figure 1. As expected, one obtains different estimates of the target accuracies, depending whether the correlations between stations are taken into account or not.

APPENDIX: Iteration Formulas

We provide a set of iteration formulas that are derived from the Newton equations (12) by algebraic manipulations. First, we define the following matrices:

$$G = (F_X R F_X^T)^{-1} \tag{A.1}$$

$$A = R F_X^T G F_X - I \tag{A.2}$$

$$\Gamma = [I + AR(K^T F)_{XX}]^{-1} \tag{A.3}$$

$$E_0 = \Gamma \cdot [AC - R F_X^T G F] \tag{A.4}$$

$$E_1 = \Gamma \cdot [R F_X^T G F_t + AR(K^T F)_{Xt}] \tag{A.5}$$

$$D_0 = (K^TF)_{tX} - F_t^TGF_XR(K^TF)_{XX} \tag{A.6}$$

$$D_1 = (K^TF)_{tt} - F_t^TGF_XR(K^TF)_{Xt} \tag{A.7}$$

$$N = F_t^TGF_t - D_1 + D_0E_1 \tag{A.8}$$

The iteration equations are

$$N\tau = F_t^TG(F_XC - F) + D_0E_0 \tag{A.9}$$

$$K + \kappa = G(F_XC - F) + G[F_t + F_XR(K^TF)_{Xt}]\tau - GF_XR(K^TF)_{XX}\varepsilon \tag{A.10}$$

$$\varepsilon = E_0 - E_1\tau \tag{A.11}$$

Numerical experiments have shown that the convergence of the iteration is enhanced if the equations are used in a subiteration mode by iterating alternatively on the parameters and residuals, respectively. For parameter subiteration only equations (A.9) and (A.10) are used, assuming $\varepsilon \equiv 0$. For residual subiteration one sets $\tau \equiv 0$ and uses equations (A.10) and (A.11).

In the variance formula (17) one uses N, defined by equation (A.8) and

$$S = F_t^TGF_X + D_0\Gamma A \tag{A.12}$$

Another equivalent set of Newton-Raphson iteration equations is given in [20]. None of the sets is numerically superior to the other, and both require subiterations of parameters and residuals for efficiency.

Gauss-Newton iteration equations can be obtained from Newton-Raphson iteration equations by setting all second order derivatives equal to zero. The convergence of Gauss-Newton algorithms is inferior, but in some applications they have a larger domain of convergence.

LIST OF SYMBOLS

c, C	Residuals
f, F	Constraint (model) functions
k, K	Correlates (Lagrange multipliers)
n	Dim X (total number of observations)
N	Coefficient matrix for variance estimation of t
p	Dim t (number of parameters)
r	Dim F (number of scalar constraint equations)
R	Estimated variance-covariance matrix of the observations X
s	Number of subsets in a partitionable problem

S	Influence matrix for variance estimation of t
t, T	Parameters
x, X	Observables
V_t	Estimated variance-covariance matrix of t
W	Objective function = $c^T R^{-1} c$
$\tilde{W}$	Modified objective function
ε	Correction of C
κ	Correction of K
τ	Correction of T

REFERENCES

1. A. Albert. *Regression and the Moore-Penrose Pseudoinverse*. Academic Press, New York, 1972.
2. Y. Bard. *Nonlinear Parameter Estimation*. Academic Press, New York, 1974.
3. P. R. Bevington. *Data Reduction and Error Analysis for the Physical Sciences*. McGraw-Hill, New York, 1969.
4. A. Bjerhammar. *Theory of Errors and Generalized Matrix Inverses*. Elsevier, Amsterdam, 1973.
5. T. L. Boullion and P. L. Odell. *Generalized Inverse Matrices*. Wiley Interscience, New York, 1971.
6. S. Brandt. *Statistical and Computational Methods in Data Analysis*. North-Holland, Amsterdam, 1970.
7. H. I. Britt and R. H. Luecke. The estimation of parameters in nonlinear, implicit models, *Technometrics*, Vol. 15, pp. 233-247 (1973).
8. D. Brown. A matrix treatment of the general problem of least squares considering correlated observations, Report BRL-937, Ballistic Research Laboratories, Aberdeen Proving Ground, Maryland (1955).
9. A. Celmiņš. Least squares adjustment with finite residuals for nonlinear constraints and partically correlated data, Report R-1658, Ballistic Research Laboratories, APG, Maryland (1973).
10. A. Celmiņš. A manual for general least squares model fitting, Report ARBRL-TR-02167, Ballistic Research Laboratories, APG, Maryland (1979).
11. C. Daniel and F. S. Wood. *Fitting Equations to Data*, 2 Ed. Wiley, New York, 1980.
12. W. E. Deming. *Statistical Adjustment of Data*. Wiley, New York, 1944.
13. N. R. Draper and H. Smith. *Applied Regression Analysis*. Wiley, New York, 1967.
14. J. R. Green and D. Margerison. *Statistical Treatment of Experimental Data*. Elsevier, Amsterdam, 1978.
15. W. C. Hamilton. *Statistics in Physical Science*. Ronald Press, New York, 1964.

16. W. H. Jefferys. On the method of least squares, *The Astronomical Journal*, Vol. 85, pp. 177-182 (1980).

17. J. J. Moré, B. S. Garbow, and K. E. Hillstrom. User Guide for MINPACK-1, Report ANL-80-74, Argonne National Laboratory, Argonne, Illinois (1980).

18. E. Niple. Nonlinear least squares analysis of atmospheris absorption spectra, *Applied Optics*, Vol. 19, pp. 3481-3490 (1980).

19. R. M. Passi. Use of nonlinear least squares in meteorological applications, *Journal of Applied Meteorology*, Vol. 16, pp. 828-832 (1977), and Vol. 17, pp. 1579-1580 (1978).

20. A. J. Pope. Two approaches to nonlinear least squares adjustments, *The Canadian Surveyor*, Vol. 28, pp. 663-669 (1974).

21. J. E. Rall and R. E. Funderlic. Interactive VARPRO (INVAR), a nonlinear least squares program, Report ORNL-CSD-55, Union Carbide Corporation, Oak Ridge, Tennessee (1980).

22. A. Ruhe and P. A. Wedin. Algorithms for separable nonlinear least squares problems, Report SU326 P30-31, Stanford University (1974).

23. J. M. Thomas, M. I. Cochran, C. R. Watson, and L. L. Eberhardt. COMP--A basic language nonlinear least squares curve fitting program, Report PNL-2409, Battelle Pacific Northwest Laboratories, Richland, Washington (1977).

24. J. M. Tienstra. *Theory of the Adjustment of Normally Distributed Observations*. N. V. Uitgeverij "Argus," Amsterdam, 1956.

25. J. R. Wolberg. *Prediction Analysis*. D. van Nostrand Company, Princeton, New Jersey, 1967.

INDEX

Algorithm acceleration by sensitivity analysis, 114-123

Badly behaved constraints, 60
Best approximations of vector-valued functions, 45-47
Block diagonal matrix structures, 141
Boundary value problems, 42-45
Bounded selection, 25

Closed point-to-set mapping, 68
Complementary slackness, 51
Cone
  of affine directions, 53
  of directions of constancy, 53
  of directions of decrease, 53
  of directions of nonincrease, 53
  linearizing, 52
  nonnegative polar, 51
  of subgradients, 52
  tangent, 52,53
Constant rank theorem, 3-6
Convex program, 50,83

Differentiable parameter functions, 36-42
Dini derivative bounds, 88
Directional derivative, 52,77
  limit quotient bounds, 77-85
  limit quotient infimal sequence, 79
Dirichlet problem, 43
Dual program, 59

Elliptic vector norm, 133
Equilibrium conditions, 92-93
  sensitivity analysis, 103-105
Extremal value function, 8
  continuity properties, 10-13

Faithfully convex function, 54
  piecewise, 57
Feasible set, 7,23,51,53,67
  point-to-set mapping
    closed, 75,76
    uniformly compact, 76

Gauss-Newton equations, 136
Geometric programming
  dual, 111
  dual harmonic, 113
  harmonic, 112,113
  primal, 110,111
  reversed, 112

Harmonic mean algorithm, 114
  sensitivity analysis modification, 114-123
Harmonic program, 112

Implicit function theorem, 2,69

Jacobian matrix
  generalized, 2
  usual, 2

Karush-Kuhn-Tucker theorem, 51
Kuhn-Tucker multipliers (*see* Kuhn-Tucker vectors)
Kuhn-Tucker vectors, 50,51,53-59,68

Lagrangian
  dual, 59
  for a general parametric NLP, 67
Least squares model example, 145-149
Linear independence assumption, 82
Linearizing cone, 52
Lipschitz function, 11
Lower semicontinuity of solution set point-to-set mapping, 25-47
  necessary conditions, 30-42
  sufficient conditions, 25-30, 36,37,43

Mangasarian-Fromovitz Constraint Qualification, 9,68,70,71
Mathematical model, 132
Maximum likelihood, 133,134
Metric projection, 45
MFCQ (*see* Mangasarian-Fromovitz Constraint Qualification)
Minimal formulation, 142
Minimal points, 23
Minimum value, 23
Model fitting problem, 132-135

Newton-Raphson equations, 135-136
Nonlinear complementarity problem, 99
Nonnegative polar cone, 51
Normal equations, 135
  sensitivity calculations, 136, 137

Open point-to-set mapping, 68
Optimal value function (*see also* Minimum value and Extremal value function and Perturbation function), 68
  continuity of, 75,76
  lower semicontinuity, 76
Optimality conditions
  first order sufficient, 16
  second order sufficient, 16,18

Parameters
  essential, 142
  nonessential, 142
Parametric NLP problems, 7,23,25, 43,50,53,67,70,83,88
Partitionability conditions for least squares model, 141
Perturbation function, 53
Perturbed feasible set, 53
Perturbed program, 53
Piecewise faithfully convex, 57
P-mapping, 25
Point-to-set mapping
  closed, 68
  lower semicontinuity, 25-47
  open, 68
  solution-set, 24
  uniformly compact, 68
Polyhedral function, 54

Rank conditions, 134
Reduced program, 70
Reduction of variables, 69-70
Regularity condition, 9
Regularizing set, 58
Regular point, 9
Restricted Lagrangian dual program, 59
Reversed geometric program, 112
  equilibrium solution, 114
  harmonic program, 112,113
  harmonic program approximation, 114
Right-hand side perturbation problem (*see also* Perturbed program), 53
  equivalent to general parametric problem, 88

Shadow prices, 50
Slater's condition, 9,51
Solution set, 24,67
  stability bounds, 14-18
Solution-set point-to-set mapping, 24
  closed, 78
  lower semicontinuity, 25-47
Solvability condition, 94,95
Spatial equilibrium conditions, 92,93
Stability of differentiable nonlinear systems, 9,10
Stability of local solution set, 14
Stable perturbations, 57,58
Stable program, 54
Standard error, 137
Standard least squares problem, 139-140
Subdifferential, 51

Totally partitionable, 140

Uniformly compact point-to-set
mapping, 68

Variance-covariance matrix
of the observations, 133,134

[Variance-covariance matrix]
of the parameters, 137

Weighted least squares, 133